Reading and Writing Strategies *for the* Secondary SCIENCE Classroom *in a* PLC at Work®

Daniel M. Argentar
Katherine A. N. Gillies
Maureen M. Rubenstein
Brian R. Wise

EDITED BY
Mark Onuscheck
Jeanne Spiller

555 North Morton Street
Bloomington, IN 47404
800.733.6786 (toll free) / 812.336.7700
FAX: 812.336.7790

email: info@SolutionTree.com
SolutionTree.com

Visit **go.SolutionTree.com/literacy** to download the free reproducibles in this book.

Printed in the United States of America

Library of Congress Cataloging-in-Publication Data

Names: Argentar, Daniel M., 1970- author.
Title: Reading and writing strategies for the secondary science classroom in a PLC at work / Daniel M. Argentar, Katherine A.N. Gillies, Maureen M. Rubenstein, Brian R. Wise, Mark Onuscheck (Series Editor), Jeanne Spiller (Series Editor).
Description: Bloomington, IN : Solution Tree Press, [2019] | Series: Every teacher is a literacy teacher series | Includes bibliographical references and index.
Identifiers: LCCN 2019010240 | ISBN 9781949539011 (perfect bound)
Subjects: LCSH: Science--Study and teaching (Middle school)--United States. | Science--Study and teaching (Secondary)--United States. | Reading (Middle school)--United States. | Reading (Secondary)--United States. | Composition (Language arts)--Study and teaching (Middle school)--United States. | Composition (Language arts)--Study and teaching (Secondary)--United States. | Interdisciplinary approach in education--United States.
Classification: LCC Q183.3.A1 A74 2019 | DDC 507.1/273--dc23 LC record available at https://lccn.loc.gov/2019010240

Solution Tree
Jeffrey C. Jones, CEO
Edmund M. Ackerman, President

Solution Tree Press
President and Publisher: Douglas M. Rife
Associate Publisher: Sarah Payne-Mills
Art Director: Rian Anderson
Managing Production Editor: Kendra Slayton
Senior Production Editors: Tara Perkins and Todd Brakke
Content Development Specialist: Amy Rubenstein
Copy Editor: Miranda Addonizio
Proofreader: Jessi Finn
Cover Designer: Rian Anderson
Editorial Assistant: Sarah Ludwig

ACKNOWLEDGMENTS

We are indebted to all of the mentors, teachers, and students who have shaped our thinking over the course of our teaching careers. In particular, we are grateful for the partnership and collaborations we have enjoyed with the Stevenson High School science department. It is their commitment to literacy within our professional learning community that is the foundation for adapting and creating many of the strategies in this book. Thanks go out to science director Steve Wood and teachers Sara Cahill, Kellie Dean, Brett Erdmann, Caroline Humes, Amy Inselberger, Kim Lubecke, Abbie Lueken, Christine Pfaffinger, Deanna Warkins, and Tommy Wolfe. Your passion, expertise, inspiration, and support have been invaluable.

Thank you to our Stevenson and Niles North High School colleagues on our academic literacy teams. At Stevenson, Christina Anker, Jim Barnabee, and Nicole Fuller all contributed to several reading and writing strategies in this book. At Niles North, all teachers who have served on our reading collaborative teams have truly modeled the hard work of collaboration as well as reflective and impactful teaching.

Thank you to D219 Niles Township High Schools' literacy coaches and specialists Stephanie Iafrate, Ellen Foley, Christine Mbah, and Mary Richards. Your commitment to fostering a culture of literacy at D219 has tremendously impacted this important body of work.

We also thank the literacy specialists, coaches, and teachers from Downers Grove North High School and the Chicago Area Literacy Leaders (CALL) group for sharing their expertise and inspiring the important literacy work and collaborations we engage in daily.

Thank you to the administrative leaders at Stevenson and Niles North for their encouragement and support. By prioritizing literacy in our schools, they have allowed our work to grow and inspire our school communities.

Finally, a very special thanks goes out to Mark Onuscheck, director of curriculum, instruction, and assessment at Stevenson High School, whose compassion, humor, inspiration, guidance, and friendship motivate and inspire us regularly. Without his steady leadership, there would be no book to write.

Solution Tree Press would like to thank the following reviewers:

Kara Cheslock
Science Teacher
Batchelor Middle School
Bloomington, Indiana

Erin Kowalik
Science Department Chair
James Bowie High School
Austin, Texas

Nicole McRee
Science Coach
Twin Groves Middle School/
Woodlawn Middle School
Buffalo Grove/Long Grove, Illinois

Kelly Melendez Loaiza
Science Teacher
Mansfield High School
Mansfield, Massachusetts

Kristen Vileta
Science Teacher
Twin Groves Middle School
Buffalo Grove, Illinois

Natalie Wilson
Science Teacher
Hudson Middle School
Sachse, Texas

Visit **go.SolutionTree.com/literacy** to download the free reproducibles in this book.

TABLE OF CONTENTS

Reproducible pages are in italics.

ABOUT THE SERIES EDITORS

Mark Onuscheck is director of curriculum, instruction, and assessment at Adlai E. Stevenson High School in Lincolnshire, Illinois. He is a former English teacher and director of communication arts. As director of curriculum, instruction, and assessment, Mark works with academic divisions around professional learning, articulation, curricular and instructional revision, evaluation, assessment, social-emotional learning, technologies, and Common Core implementation. He is also an adjunct professor at DePaul University.

Mark was awarded the Quality Matters Star Rating for his work in online teaching. He helps to build curriculum and instructional practices for TimeLine Theatre's arts integration program for Chicago Public Schools. Additionally, he is a National Endowment for the Humanities' grant recipient and a member of the Association for Supervision and Curriculum Development, the National Council of Teachers of English, and Learning Forward.

Mark earned a bachelor's degree in English and classical studies from Allegheny College and a master's degree in teaching English from the University of Pittsburgh.

Jeanne Spiller is assistant superintendent for teaching and learning for Kildeer Countryside Community Consolidated School District 96 in Buffalo Grove, Illinois. School District 96 is recognized on AllThingsPLC (www.AllThingsPLC.info) as one of only a small number of school districts where all schools in the district earn the distinction of a model professional learning community. Jeanne's work focuses on standards-aligned instruction and assessment practices. She supports schools and districts across the United States to gain clarity about and implement the four critical questions of professional learning communities. She is passionate about collaborating with schools to develop systems for teaching and learning that keep the focus on student results and helping teachers determine how to approach instruction so that all students learn at high levels.

Jeanne received a 2014 Illinois Those Who Excel Award for significant contributions to the state's public and nonpublic elementary schools in administration. She is a graduate of the 2008 Learning Forward Academy, where she learned how to plan and implement professional learning that improves educator practice and increases student achievement. She has served as a classroom teacher, team leader, middle school administrator, and director of professional learning.

Jeanne earned a master's degree in educational teaching and leadership from Saint Xavier University, a master's degree in educational administration from Loyola University, Chicago, and an educational administrative superintendent endorsement from Northern Illinois University.

To learn more about Jeanne's work, visit www.livingtheplclife.com, and follow @jeeneemarie on Twitter.

To book Mark Onuscheck or Jeanne Spiller for professional development, contact pd@SolutionTree.com.

ABOUT THE AUTHORS

Daniel M. Argentar is a literacy coach and communication arts teacher at Adlai E. Stevenson High School in Lincolnshire, Illinois. As a sixth-grade teacher, he taught reading, language arts, social studies, and science. Since 2001, he has provided academic literacy support to struggling freshmen and sophomores, in addition to teaching other college prep and accelerated English courses. In his coaching role, he partners with instructors from all divisions to increase disciplinary literacy for students—running book studies, professional development sessions, and one-on-one coaching meetings.

Daniel received a bachelor's degree in speech communications from the University of Illinois at Urbana-Champaign, an English teaching degree and a master's degree in curriculum and instruction from Northeastern Illinois University in Chicago, and a master's degree in reading from Concordia University in Chicago.

To learn more about Daniel's work, follow @dargentar125 on Twitter.

Katherine A. N. Gillies works as a reading specialist and English teacher at Niles North High School in Skokie, Illinois, where she previously served as a literacy coach. Here, Katherine serves as the lead architect of schoolwide literacy improvement work, including building a comprehensive system of intervention and support for struggling readers as well as crafting research-based curricula to ensure the continued literacy growth of all students. Katherine leads a number of collaborative teams and cross-curricular initiatives aimed at using data to inform instruction, building capacity for disciplinary literacy, and employing responsible assessment practices in the secondary arena. She has presented at local and national conferences, including that of the National Council of Teachers of English, on these topics.

Katherine earned a bachelor's degree in literature and secondary education from Saint Louis University; a master's degree in literacy, language, and culture with reading specialist certification from the University of Illinois, Chicago; and a master's degree in educational leadership and administration from Concordia University, Chicago. She is also a certified trainer for Project CRISS (Creating Independence through Student-Owned Strategies).

To learn more about Katherine's work, follow @Literacyskills on Twitter.

Maureen M. Rubenstein is a literacy coach and special education instructor at Adlai E. Stevenson High School. As a teacher, she works with students on individualized education plans who have diagnosed reading, writing, and emotional disabilities. In her coaching role, she partners with instructors from all divisions to work on disciplinary literacy. In addition to coaching individual teachers, she works with the other literacy coaches to coordinate and implement book clubs, professional development sessions, and one-on-one coaching sessions.

Maureen received a bachelor's degree in special education from Illinois State University, a master's degree in language literacy and specialized instruction (reading specialist) from DePaul University, and a master's degree in educational leadership from Northern Illinois University. Maureen is also a certified trainer for Project CRISS, and she is certified to teach Wilson Reading.

To learn more about Maureen's work, follow @SHS_LiteracyMR on Twitter.

Brian R. Wise is a literacy coach and English teacher at Adlai E. Stevenson High School in Lincolnshire, Illinois. He has taught a wide array of English and literacy intervention courses throughout his teaching career. As a literacy coach, he works with faculty members from all divisions of the high school to build teachers' capacity for embedding literacy skills into classroom instruction and assessment.

Brian received his bachelor's degree in English education from Boston University, a master's degree in English from DePaul University, and master's degrees in reading and in principal leadership from Concordia University, Chicago.

To learn more about Brian's work, follow @Wise_Literacy on Twitter.

To book Daniel M. Argentar, Katherine A. N. Gillies, Maureen M. Rubenstein, or Brian R. Wise for professional development, contact pd@SolutionTree.com.

PREFACE

To begin this book, and to immediately demonstrate the value of professional learning communities (PLCs) to support positive, thoughtful collaboration, we want to share a real-life experience we had with a group of fellow teachers in our school. We believe this serves as an example of the familiar struggle occurring in many schools when teachers from various content areas struggle to approach literacy instruction.

"I don't have time for literacy" and "I don't know how to teach literacy" are two challenges we often hear when we work with teachers on how to integrate literacy strategies into their classrooms in ways that support learning. In the fall of 2013, we, the authors of this book serving together as a group of literacy coaches, were asked to present a workshop about science literacy to a group of science teachers working to adopt the new Next Generation Science Standards (NGSS; NGSS Lead States, 2013). Our purpose was to introduce how the NGSS were designed to focus teaching and learning practices on argumentation and literacy within science curricula. Our goals were not only to be a resource for our science teachers as they began to align their curriculum to the NGSS for grades 6–12, but also to help them better understand how a focus on literacy supports their efforts to teach students to think like scientists. At that time, many of those science teachers recognized the need to elevate student literacy skills, but many of them felt underprepared in this area. We couldn't blame them. After all, few were trained to teach literacy, and almost all of them felt stressed about the amount of science content they had to cover to prepare their students for the future. On top of everything else, trying to find time to integrate literacy strategies into teaching and learning seemed overwhelming to them.

As literacy coaches, we spent three days that fall collaborating with teams of science teachers, setting goals that focused on results, and generating innovative approaches to student learning. As we patiently offered ideas and listened as these teachers developed their plans, we learned about the challenges they faced in teaching science, the essential skills they planned to focus on, and the ways they discussed student learning in science. As we began to build collaborative relationships with our science teachers, we began to seed our discussions with ideas about literacy, and we worked with them to consider how a focus on literacy could connect with their focus on important science standards and the skills needed to develop as a scientist.

However, as they revised their curriculum, their focus on the content skills constantly outweighed their focus on the literacy skills necessary for students to become strong science learners. Concerns about cutting into content learning time and about the lack of ability to teach literacy came up again and again. Some teachers simply said, "I don't have time for literacy." Some said, "I need to focus more on the science." We soon concluded that a three-day workshop was not enough to truly integrate the changes we were hoping to make.

We knew we had more work to do, and we knew we would need to rethink how we could support teachers—not just through a short workshop. As we rethought our approach, we continued to return to our values and the commitments we had made to our PLC. In so doing, we considered ways to design our literacy work by establishing a focus committee dedicated to building literacy-based strategies and developing ongoing literacy coaching for our teachers who wanted to think differently about teaching and learning. By taking these steps, and with the support of our administrators and school district, we worked to make use of our PLC culture to impact change.

Although it was a struggle for the whole group to digest at one time, our initial introduction to literacy and the NGSS ended up resonating with some of the teachers. This planted literacy seeds that we were able to pursue with a smaller group of science teachers. The story of Cami (a pseudonym we use to ensure privacy) highlights our successful process of engaging these teachers and inspiring them to volunteer to join us (the literacy coaches) and to join our schoolwide literacy committee.

Early in the year, after our initial presentations on science and literacy, we had an encounter with a science teacher and colleague named Cami. We knew that, as a veteran teacher, Cami was a strong voice in her department, and we wanted to convince her to join our literacy committee so we could support her work and help her students progress in her science classroom. We talked with her about the cool things teachers could accomplish by using literacy and science content to work on targets aligned with the NGSS. We thought she was hooked—but when we formally asked her to join the committee, she told us that she knew reading was valuable, but she couldn't find the time in her lessons and schedule to integrate literacy skills alongside the science content skills. She told us, "There is just so much curriculum to get through. I don't think I have time to try and teach literacy alongside teaching the science."

Later in the year, we renewed our efforts to gain access to Cami's science classroom. We set up another meeting with her to try to convince her about the value of focusing on literacy in the science classroom, but this time, we brought her data about her students that demonstrated the range of reading abilities in her classroom. As we went through these data, she began to notice patterns that supported her concerns for certain students who seemed to struggle in her science classroom. Throughout the meeting, we reminded her of some of the great ideas we had chatted about earlier in the year, and we spoke about the need to better prepare our students for college and career readiness. Again, Cami was happy to talk, but when push came to shove, she reminded us she had "no time to teach literacy."

Then, one day later that spring, Cami reached her breaking point. Faced with students' consistent struggles to perform well on assessment questions that were reading dependent, students' continued reliance on her lectures and explanations and their lack of comprehension of the text, and students' apparent refusal to do most textbook reading, Cami came to us and asked, frustrated, "Okay, I give up. How do I get them to read?"

She knew something needed to change, and we were eager to help. During that short meeting, we focused Cami on our purpose in working with teachers in different academic disciplines. We replied, "You need to show your students how *you* think—how a scientist thinks—and why it's important for a scientist to consume and use information from texts." That exchange was the start of a now-long-standing collaborative relationship about literacy, science, and student learning.

Cami's frustration with trying to hold students accountable for reading and writing turned into a collaborative experience combining content and literacy learning in a science classroom. Together, we planned a sequence of lessons that did not center on lecturing but instead would use strong literacy strategies to help students think about what information is important for scientists, comprehend the text they read, and synthesize information in ways that would increase their ability to master higher-order science standards. We collaborated to generate many different ways to approach literacy in the science classroom that continuously supported Cami's focus on the skills of the science curriculum. She began to see how literacy skills and higher-order-thinking skills interconnect, and we worked together to help integrate literacy skills that supported how to learn science. Through this work, we built a collaborative partnership that continues to be more and more innovative.

Through our collaboration, Cami learned how to model ways a scientist would approach reading, how to use think-aloud strategies, how to guide students through the reading process, and how to support students in assessing their reading performance. Applying these strategies led her students to use the information from the text actively in class instead of passively listening to a lecture. She was thrilled with her students' engagement and improving performance. At the end of the school year, when we sent out our invitation for teachers to join our literacy committee, Cami committed, and so did many other teachers who were seeking to make positive changes in their science classrooms. The next fall, not only did we continue our collaborative work with Cami, but we also continued to innovate new ways of working with an expanding group of teachers dedicated to our literacy committee and one-on-one literacy coaching in the science classroom.

As is our practice, the first thing we do with any new teacher to our literacy committee is sit down to have a conversation. The conversation can be formal or informal, but the purpose is to get to know the teacher, to become familiar with the types of students the teacher generally teaches, to explore any initial literacy concerns the teacher might already have coming into our work, and to set some collaborative goals to focus our work. Obviously, by then we knew Cami pretty well, but the issues she identified at that initial meeting with her resonated with many of the science teachers we worked with. During the collaborative work between literacy coaches and science teachers, the following issues surfaced regularly throughout our conversations.

- Fears of teaching reading among teachers because they did not have training as literacy teachers
- Concerns about teachers' losing time to teach content while also teaching literacy
- Struggles among students to see a clear purpose for reading and to synthesize information after they finish a reading assignment
- A need for focused reading strategies to help students be successful
- A need to support students using information and applying or connecting reading to scientific problems

- Struggles among students with analyzing visual information, such as graphics and data tables
- A need for students to improve the way they write lab conclusions, including a need to improve focus, increase the use of evidence, and provide clearer justification
- A need for students to broaden and increase their knowledge of science with choice reading opportunities within and outside the current unit of instruction
- Concerns that some students were faking their way through reading or avoiding it altogether (*pseudo-reading*; Buehl, 2017)

For all of us, from literacy coaches to science teachers, the initial goal was clear—we needed to equip our teachers with the ability to help improve their students' reading skills to address the expectations of the NGSS and to better prepare them for the reading skills they would require for college and career readiness.

Our work began with developing an understanding of how scientists consume and use information—the disciplinary literacy of science—and Cami's commitment has since led to several years of creative collaboration aimed at teaching students to think like scientists and to use literacy within the NGSS.

We share this story from the start because it exemplifies three important commitments that are core to our work and the work of a strong PLC: (1) we believe in supporting collaboration between experts seeking to solve an educational concern; (2) we believe in integrating change that is focused on every student's learning—where teams systematically consider, implement, evaluate, and revise all changes; and (3) we believe that we must focus on the results students produce.

This book is dedicated to the literacy issues we think are important to pay attention to when connecting literacy to science. We hope these ideas can help develop collaborative partnerships at your school, and we hope this book can serve as a strong resource for your teaching if a literacy expert isn't an immediate resource for your work in teaching and learning. Use this book as a thought partner, and make literacy a priority. Every teacher is a teacher of literacy.

INTRODUCTION

Every Teacher Is a Literacy Teacher

In this series of books, called *Every Teacher Is a Literacy Teacher*, we focus on how each subject area in the grades 6–12 experience has a need to approach literacy in varying and innovative ways. To address this need, we designed each book in the series to:

- Recognize the role every teacher must play in supporting the literacy development of students in all subject areas throughout their schooling
- Provide commonly shared approaches to literacy that can help students develop stronger, more skillful habits of learning
- Demonstrate how teachers can and should adapt literacy skills to support specific subject areas
- Model how a commitment to a PLC culture can promote the innovative collaboration necessary to support the literacy growth and success of every student
- Focus on creating literacy-based strategies in ways that promote the development of students' critical-thinking skills in each academic area

You may immediately recognize how this approach differs from many traditional school practices and formats, where educators view literacy development as the job of English language arts (ELA) teachers, reading teachers, or teachers of English learners. It is an accepted practice that these teachers bear the responsibility of teaching skills like vocabulary development, comprehension skills, inferential skills, and writing skills. In stand-alone ways, they shoulder the charge to single-handedly support the literacy growth of students. While these teachers and traditional education approaches may be effective to a degree, we recognize the need for our schools to support changes that make teaching literacy a responsibility for all teachers. In this book, which focuses on the science classroom, we propose that schools adopt the collective commitment that every teacher is a literacy teacher. This commitment means we must support collaboration between expert science teachers and experts in literacy using processes similar to in the story of Cami that we detailed in this book's preface (page xvi).

As we begin to aggressively address literacy issues in our classrooms, PLCs need to recognize the value of supporting literacy skills within every classroom—and every content area. Science teachers need to be literacy teachers. Mathematics teachers need to be literacy teachers. Social studies teachers need to be literacy teachers. World language and fine arts teachers need to be literacy teachers. Every teacher needs to be a literacy teacher. By making literacy a core commitment in the work of every academic discipline, schools can begin to develop students' abilities to read and write with a variety of more focused literacy strategies that support the critical-thinking skills necessary for science, social studies, mathematics, language acquisition, and the fine arts.

In this book, we emphasize how building collaboration among science teachers and literacy experts will be one of our greatest catalysts for supporting student growth in every area of school curriculum, and we stress a strong commitment toward building instructional improvements that can support the growth of every learner. As we've seen in many PLC cultures, collaboration generally begins with teaming teachers within like disciplines. Science teachers team with other science teachers, social studies teachers team with other social studies teachers, and so on. When teams form according to discipline, they tend to focus only on their content and discipline-based skills. We intend for this book to encourage collaboration of a different sort—collaboration among literacy and science experts teaching middle school and high school. When discipline-based teachers and literacy experts team up, they can build stronger approaches to teaching and learning that connect literacy-based strategies with discipline-specific subject areas. When these two forces come together to collaborate, we begin to see positive results.

As we constructed this book, we recognized that many schools do not have *literacy experts* (English teachers, reading teachers, reading specialists, and so on) available to collaborate with science teachers around the challenges of building stronger scientific readers and writers. To that end, we encourage you to use this book as a thought partner with your team or as your own personal literacy expert that can help you generate changes to support student learning. In either case, we mean for this book to be a helpful companion as you deepen conversations and navigate choices that will positively affect student growth and development, and we structured the text to demonstrate how to not only develop collaborative practices but also support both individual readers and teams in becoming reflective practitioners.

As you will see, this book provides, describes, and details many literacy-based strategies that you can integrate into the science classroom. You can use many of the strategies immediately; others require preparation. In either case, we highly encourage getting started. Integrating focused literacy strategies into the science classroom initiates and promotes significant gains in learning, deep comprehension, and the capacity to think critically.

There are many reasons why science teachers in grades 6–12 need to be literacy teachers. Reading about science, writing about science, and thinking like a scientist require a mindset that focuses on elements of reading and writing that are fundamentally different from reading fiction or history or the news. Reading and writing about science requires the following.

- A close attention to detail
- An understanding of how details interconnect to build conceptual understandings
- The ability to interpret and synthesize data

Literacy strategies create an infrastructure of supports that allow students to enter into science, rather than do an exercise in memorization and information recall. Instead, stronger literacy strategies provide the necessary skills that support students' abilities to think like scientists with prereading, during-reading, and post-reading experiences that are interconnected to the demands of becoming a young scientist. We believe it is necessary to make use of literacy strategies in a way that supports the thinking of science, and there are many innovative, engaging ways to support that commitment.

The Need for Literacy Instruction

Picture a reader who is just beginning to learn how to read. What behaviors do you see as this student engages with text? What is he or she learning to do first? How is he or she grappling with the challenge of learning how to read? Chances are, you visualize this reader at the beginning stages, working to crack the alphabetic code—breaking apart and sounding out words, one syllable at a time, and likely dealing with simple language and colorful text. The words the student is trying to read are already ones that he or she likely employs in conversation. This student is engaging in growing basic literacy skills—decoding, fluency, and automaticity. During this early phase of learning how to read, comprehension and meaning making almost take a back seat to decoding. The reader is working on the mechanical process of learning to read.

As readers advance beyond the beginning stages of reading and advance in their abilities to read, they become more fluent and able to comprehend a text. At this point, the advanced reader possesses the ability to make meaning from what he or she reads—the process of reading is no longer dedicated to the mechanical process of encoding and decoding a text. Instead, the process of reading is dedicated to learning and thinking. More advanced readers are able to infer from and analyze what they read in a book, as well as what they read in the world, even when they have limited experience with a topic. Such readers possess the critical literacy skills they will need for college and success in the workplace. These critically literate students are ready to take on complex tasks and dive into disciplinary literacy tasks—tasks that are specific to particular subject areas like science.

Now, what about the reader who is somewhere between these two phases—the reader who is not a beginning reader and is not an advanced reader? What about the student who can break the code—he or she can encode and decode—but struggles to apply this information to make new understandings? The reality that we all know and experience in our classrooms is that there are many students who fall into this place along the continuum, and there are many students who leave our high schools without the essential life skill of being critically literate. In fact, National Assessment of Educational Progress (NAEP) results detailed in *The Condition of Education 2018* report (McFarland et al., 2018) suggest that only 36 percent of eighth-grade students and 37 percent of twelfth-grade students possess literacy skills at or above the level of proficiency and over 60 percent have not met this readiness benchmark. This means that a majority of students are moving through middle school and high school without developing the literacy skills necessary to be successful in science classrooms. This is the group of students with which we are most concerned in this book. We know that this large group of students requires greater attention and a greater concentration on skill development. Moreover, a specific portion of these students will continue to need support in even basic literacy skill development. It is this portion of our student population that seems to be the conundrum—often these are the students teachers struggle to support.

A science curriculum is often incredibly challenging for students who struggle with their developing literacy skills. Unfortunately, the struggle among many of this group of students is not always transparent even though they make up the majority of students in American classrooms. The graph in figure I.1 (page 6) represents the increasing gap in literacy as students grow up within schools, boldly demonstrating the challenges we must work to solve as educators in schools. In our PLCs, we must all shoulder the responsibility of student literacy and address these alarming statistics.

Research confirms there is a real need for disciplinary literacy instruction in the science classroom. Timothy and Cynthia Shanahan (2008) note the following.

- Adolescents in the first quarter of the 21st century read no better—and perhaps worse—than the generations before them.
- For many students, the rate of growth toward college readiness actually decreases as students move from eighth to twelfth grade.
- American fifteen-year-olds perform worse than their peers from fourteen other countries.
- Disciplinary literacy is an essential component of economic and social participation.
- Middle and high school students need ongoing literacy instruction because early childhood and elementary instruction do not correlate to later success.

Among the many concerns within collaborative discussions about teaching and learning, literacy continually ranks as one of the most worrisome. In many of our discussions with teachers throughout North America, teachers across academic disciplines express three running concerns: (1) many students struggle with basic literacy skills, (2) many students read and write below grade level, and (3) many students do not know how to complete reading or writing assignments.

Gaps in literacy skills are staggering, and these gaps affect all areas of many students' education. As students are marched through their schooling, the statistics demonstrate that gaps in literacy increase over the course of many students' elementary, middle, and high school years. Columbia University Teachers College (2005) reports many students find themselves reading three to six grade levels

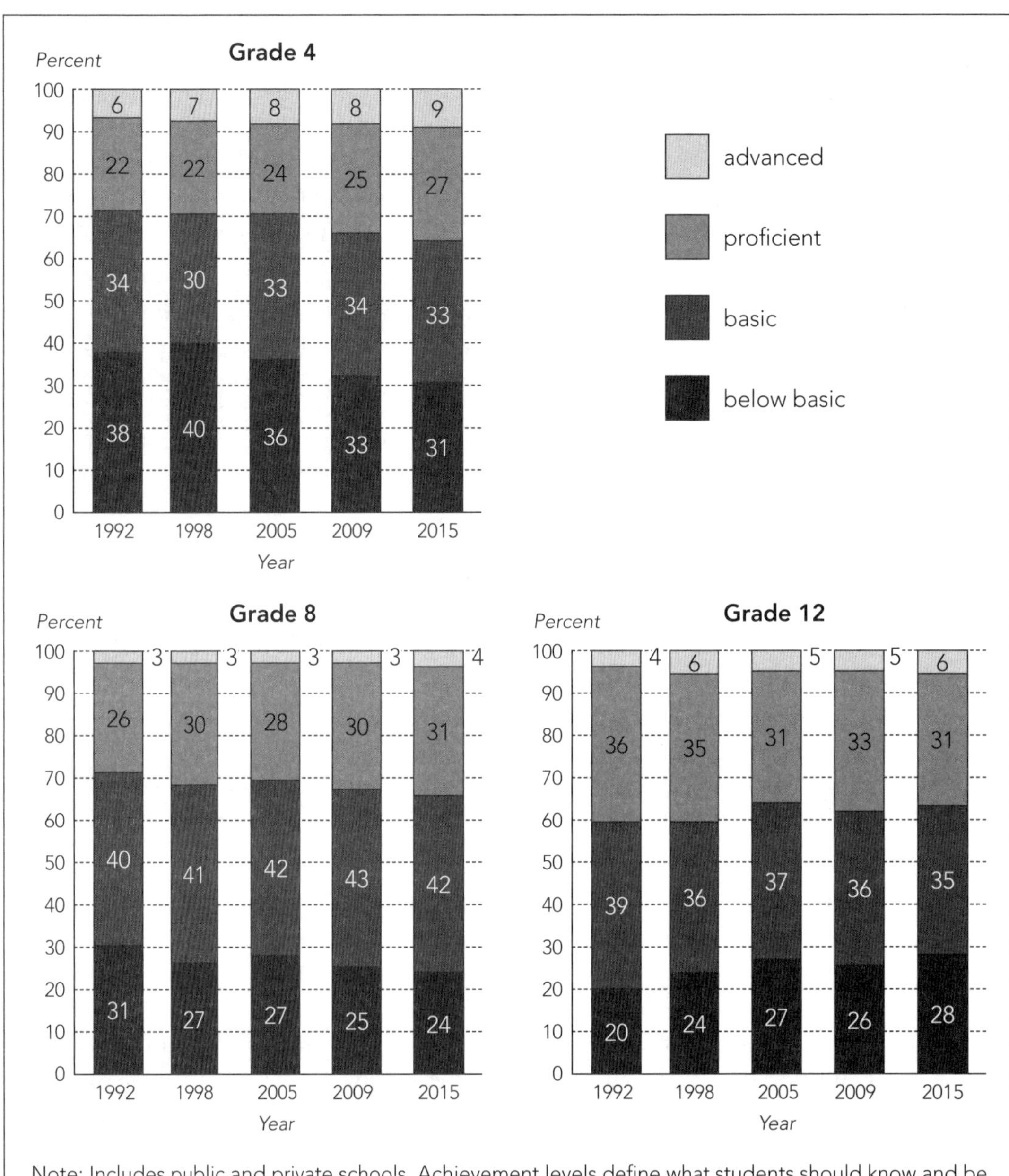

Note: Includes public and private schools. Achievement levels define what students should know and be able to do: Basic indicates partial mastery of fundamental skills, and Proficient indicates demonstrated competency over challenging subject matter. . . . Testing accommodations (such as extended time, small-group testing) for children with disabilities and English language learners were not permitted in 1992 and 1994. Although rounded numbers are displayed, the figures are based on unrounded estimates. Detail may not sum to totals because of rounding.

Source: U.S. Department of Education, National Center for Education Statistics, National Assessment of Educational Progress, n.d.

Figure I.1: Percentage distribution of fourth-, eighth-, and twelfth-grade students across NAEP reading achievement levels.

below their peers, many students struggle mightily to comprehend informational texts, and many students graduate from high school unprepared to enter a college-level experience. Columbia University Teachers College (2005) and Michael A. Rebell (2008) further highlight the following statistics, which present significant and long-standing concerns.

- By age three, children of professionals have vocabularies that are nearly 50 percent greater than those of working-class children, and twice as large as those of children whose families are on welfare.
- By the end of fourth grade, African American, Hispanic, and poor students of all races are two years behind their wealthier, predominantly white peers in reading and mathematics. By eighth grade, they have slipped three years behind, and by twelfth grade, four years behind.
- Only one in fifty Hispanic and African American seventeen-year-olds can read and gain information from a specialized text (such as the science section of a newspaper) compared to about one in twelve white students.
- By the end of high school, African American and Hispanic students' reading and mathematics skills are roughly the same as those of white students in the eighth grade.
- Among eighteen- to twenty-four-year-olds, about 90 percent of whites have either completed high school or earned a GED. Among African Americans, the rate is 81 percent; among Hispanics, 63 percent.
- African American students are only about 50 percent as likely (and Hispanics about 33 percent as likely) as white students to earn a bachelor's degree by age twenty-nine.

Statistical results like these are a stark reminder that we need to focus our attention on the literacy development of students in every corner of our schools. For the grades 6–12 science teacher, the focus on developing students' abilities to access informational texts should stand out as an important goal, as it is central to reaching science standards, building skills, meeting expectations, and developing young scientists.

In this book, we offer suggestions focused on developing intermediate literacy skills that include building academic vocabulary, self-monitoring comprehension, and knowing how to apply fix-it strategies in order to navigate a text with understanding and the ability to apply this knowledge to a prompted task (Buehl, 2017).

These are important skills to attain because students with strong intermediate literacy skills have essentially developed an awareness of their own active comprehension, and they know what to do when comprehension begins to feel shaky. It is vital that, within our disciplines, we don't jump ahead of intermediate literacy, but instead continually model this phase for our students and provide opportunities for them to practice these skills in a constructive and guided manner.

Due to its focus on literacy in the science classroom, in this book, we regularly refer to the NGSS (Next Generation Science Standards) and the CCSS ELA (Common Core State Standards for English language arts) that help to articulate the priorities teachers should support in their classrooms. In doing so, we strive to point out the interdependent relationship between literacy skills and the ability to think critically like a scientist.

Disciplinary Literacy

As you gain confidence that students have a good grasp of basic, foundational literacy skills, and as you begin to see them develop more intermediate and advanced literacy skills, you can move forward with tailoring their literacy instruction with an eye toward disciplinary literacy. Even though students will need you to continue modeling the use of academic vocabulary and monitoring their comprehension, they will also be ready to attack complex texts with a disciplinary lens even as they practice building their skills. Who better to lead the way with disciplinary thinking than the experts—our science teachers?

For our purposes, a *discipline* is a unique expertise—which schools often split into subject-matter divisions such as mathematics, science, ELA, physical education, world languages, fine arts, and so on. Disciplinary literacy focuses on the literacy strategies tailored to a particular academic subject area. This book, as previously noted, focuses on the expertise of science teachers who see the value of integrating literacy strategies into their classrooms.

What would happen if your team were to gather teachers from every discipline in your school and track the way they each address a reading, writing, and speaking task? Predict how different content-area teachers would approach and work through literacy tasks. What similarities and differences would your team observe among these varied disciplines?

Because teachers have unique expertise related to their academic field of interest, there are often noticeable differences in ways they might approach literacy-based tasks. Those differences stem from the diverse sets of expertise, interests, and background knowledge professionals each bring to teaching and learning. For each discipline in a grades 6–12 middle school or high school, teachers often attend to literacy tasks differently based on that expertise. After all, when ELA teachers read, write, and speak, they do so with certain goals and objectives in mind, such as determining universal themes, the meaning of symbols, and the author's purpose, to name a few. Those literacy goals are different in science.

There are certain stylistic and conceptual norms professionals attend to in each discipline. A scientist, a historian, a businessperson, or any other professional is going to address literacy tasks with norms and behaviors befitting his or her expertise and profession. That makes total sense; after all, each expert or professional has unique insider knowledge. Insiders have more background knowledge, subject-related vocabulary knowledge, and subject-related purpose than others without such dispositions. On the other hand, disciplinary outsiders lack sufficient background knowledge and vocabulary to navigate a disciplinary text successfully. Literacy expert Doug Buehl (2017) suggests that our job as educators is to teach students how to think like we do—as disciplinary insiders. So, unlike an English insider, a science insider approaches reading tasks with specific goals and objectives, such as locating causes and effects, finding meaningful data, analyzing experimental conclusions, and drawing connections to scientific concepts.

Text comprehension in all disciplines generally follows a similar nine-step process, illustrated in figure I.2 (page 10), but the ins and outs of application, connection, and extension reside within the specific lens of the disciplinary expert and must be modeled accordingly. Years ago, when training our peer tutors how to help struggling readers navigate disciplinary texts, Katherine Gillies crafted this poster as a guide to moving toward text comprehension.

Given the difference between disciplinary insiders and outsiders, it makes little sense that they teach students to read and write with the same general strategies and moves. After all, if we know that each school content area has its own thinking style, it makes sense that we support students to consume and produce texts with the same unique thinking style required of each content. Even students who have a solid foundation of general strategies may struggle with the specific demands of disciplinary texts. Instead of using generic strategies in every class and across the school, providing students with a varied strategy toolbox to meet disciplinary

Did I . . . ?	Strategic Comprehension Step	Before, During, or After Reading
☐	**Preview text**, ask questions, and make predictions.	**Before:** Focus and get ready to read.
☐	**Recall** what you already know about the topic.	
☐	Set a **purpose** for reading.	
☐	Make a **note-taking plan** for remembering what's important.	
☐	Define **key concepts** and important vocabulary whenever possible.	
☐	Keep your **purpose** for reading in mind.	**During:** Stay mentally active.
☐	**Make meaning** by: • Asking questions • Putting the main ideas into your own words • Visualizing what you read • Making notes to remember what's important • Making connections between the text and people, places, things, or ideas	
☐	**Be aware** of what's happening in your mind as you read. Consider: • Am I focused or distracted? • Do I need to go back to a part I didn't get and reread it? • What are my reactions to what I am reading?	
☐	**Reflect on** what you've read. Consider: • Did I find out what I needed or wanted to know? • Can I summarize the main ideas and important details in my own words? • Can I apply what I have learned? • Can I talk about or write about what I have learned?	**After:** Check for understanding.

Source: © 2019 Katherine Gillies. Adapted with permission.

Figure I.2: Reading-comprehension process checklist.

Visit ***go.SolutionTree.com/literacy*** *for a free reproducible version of this figure.*

demands will better equip them as disciplinary insiders to read like scientists, historians, and so on (Gabriel & Wenz, 2017).

Over time, we've made positive strides toward building disciplinary literacy strategies that support learning in more directed, focused, and attentive ways. We've learned that we should apply more specific strategies to different disciplines in ways that help support learning. When we speak of this shift to disciplinary literacy and training students to be insiders, what we intend to do is teach students to think differently in each classroom they encounter during their day. This is the goal of disciplinary literacy and why we often ask teachers who wonder how to teach a text, "How would you, as an expert, address the task?" As they think through their own processes, often a strategy or a focus emerges that is unique to their discipline, which allows us to help teachers recognize the value of thinking about their discipline in relation to literacy.

About This Book

Our goal for this book is to support collaborative partnerships in schools to address science teachers' literacy concerns and better equip them with ways to support their work in science classrooms. We aim to connect that work with literacy strategies to develop students' understanding and skills as they read and write about science and learn to think like scientists.

Scope

This book is designed to help literacy leaders collaborate and build literacy capacities in the middle school and high school environment. In elementary school, teachers work hard to teach students *to read.* In middle school and high school, the goal is to teach students to read *to learn*. There's a big difference between the two approaches. Moreover, as science teachers, we want reading and writing tasks to promote students' abilities not only to learn about science but to actually do science.

As we work to approach these challenges, it is very important that readers of this book recognize that each school is unique, and each student is unique—there is no one-size-fits-all pathway to literacy development. Within this book, there is a continuum of supports related to the varying needs of each school and each classroom. Sometimes we might require short-term, immediate literacy triage; sometimes long-term, sustained collaborative development; or sometimes both triage and sustained literacy-based professional development. We recognize that strong, consistently applied literacy strategies can and will help all readers develop their potential. We invite you to adapt the strategies we offer in this book to your unique needs. Many of the same literacy strategies for less complex literacy tasks still apply to more complex tasks—the only difference is the difficulty level.

Common Language

For the purposes of this book, we recognize that we need to have a common understanding of literacy and a common language around literacy development—let's not get confused by education jargon. For instance, we use the word *text* to mean a reading, an article, a chart, a diagram, a cartoon, a source of media, and so on. There are many texts we ask students to read, and please know they can be in many formats. In addition, the term *literacy leader* can be applied to a variety of educational roles. Throughout the book, a literacy leader can be anyone in your building, such as an administrator, teacher leader, reading specialist, or literacy coach. A literacy leader is someone who has a knowledge base around literacy and wants to improve the overall literacy skills of a school environment or institution. If you don't have a literacy leader at your school, don't let that stop you. Remember, you can use this book as a thought partner. The overriding message of this book is to get started with the demanding challenges of literacy that need to be tackled now, with or without a literacy coach or a school literacy leader championing the work. Any teacher and team of teachers can initiate the changes that are necessary to support student learning; this book is meant to guide you and help you understand how to approach these changes in teaching practices.

In this book, you will also often use the term *professional learning community*. A PLC is "an ongoing process in which educators work collaboratively in recurring cycles of collective inquiry and action research to achieve better results for the students they serve" (DuFour, DuFour, Eaker, Many, & Mattos, 2016, p. 10). A PLC consists of a whole-building or whole-district culture of collaboration. We believe that a commitment to collaboration can help to support and innovate literacy in every classroom, and we believe that PLC cultures promote changes that will effectively support all students.

Within a PLC culture, collaborative teams meet on a consistent basis to build innovative practices concentrated on student growth and learning. We will use the term *team* throughout the book with the understanding that all PLC teams are interdependent and are professionally committed to continuous improvement. We know that teams may look different from building to building, and we know that schools need to configure them differently based on building resources. In this book, we will refer to science teams who are collaborating in focused ways to address literacy concerns for student learning in their science classrooms.

Chapter Contents

In chapter 1, we lay out fundamental aspects of collaborative work to address teaching literacy within the science content area. In chapter 2, we will begin with more in-depth discussions about foundational literacy and many immediate interventions for literacy difficulties that require a fast solution. We call this *literacy triage*. From there, we will focus on disciplinary literacy collaboration for prereading, during reading, and postreading in chapters 3 through 5, respectively. Within these chapters of the book, we slow down intentionally to support a deeper, focused approach. We offer classroom strategies that are the result of collaborative explorations by literacy leaders and content-area teachers—providing clarity around how varying perspectives inform instruction. For each example, we will discuss the strategy's purpose, application, and literacy focus. There are also adaptations for each strategy, which include modifications for students who qualify for special education, English learners (ELs), and those who demonstrate high proficiency and can benefit from more demanding work. Note that although these first two subgroups have different needs and different reasons they might face increased challenges with learning material (such as language barriers versus developmental barriers), we group these adaptations together because we find they often serve the learning of both subgroups equally well (just for different reasons). Indeed, even though these adaptations are geared toward these subgroups, they are applicable to any students who would benefit from a variation of a strategy that serves to scaffold learning in the short term to build out long-term proficiency. Chapter 6 offers guidance for teaching writing in science. Finally, chapter 7 covers ideas for formative and summative assessment and feedback.

Throughout this text, there are opportunities for Thinking Breaks (the first of which appeared on page 8). We intend for these to help you reflect on current practices, challenges, and opportunities for growth in working with science literacy. We know that you might do this naturally, but these are the points where we think it is important to slow down and consider ways to apply the strategies we are suggesting for your own students. In addition, there will be other opportunities for Collaborative Considerations for Teams. These are chances for teams to discuss, collaborate on, or implement disciplinary literacy ideas at the end of each chapter. You and your team can use these tasks to build science literacy into your practices in more directed ways as you target your specific grade-level curriculum.

Ultimately, we hope this book is not only a resource for ideas you can implement immediately in the classroom but also a source of inspiration for collaborative opportunities between literacy leaders and content-area instructors to build literacy capacity in your building (or buildings).

thinking BREAK

As you are reading and using this book as a resource to support your teaching, what do you want to get out of the content?

Note these three considerations for your team: (1) use this book as a book study, (2) break the book down chapter by chapter and focus on specific changes, and (3) prioritize your concerns for student learning and how to best support the literacy development of your science students.

Wrapping Up

Building collaborative teams focused on literacy development can be challenging. We know you are extremely busy and have enormous amounts of content to cover, so you may be reluctant to add another layer to your already demanding workload. However, given the data that show more than half of U.S. twelfth graders graduate high school without preparation for advanced critical thinking, we must pause and consider what we are all doing as educators to better prepare students for the future. Providing students with important intermediate literacy and science disciplinary literacy skills is an important step toward building literacy proficiency.

Collaborative Considerations *for* Teams

- What are some of the unique features of science texts?
- What are some things that science experts look for when they read?
- What deficiencies do you notice in your students that might be obstacles to their understanding of your content?
- How might your team provide experiences and vocabulary to move disciplinary outsiders closer to being insiders?

CHAPTER 1

Collaboration, Learning, and Results

It is no secret that educators are busy, so convincing content-area teachers that literacy is valuable is only one step on the way to improving literacy at a school. Once we were able to convince science teachers like Cami, the teacher we introduced in the preface (see page xvi), that we all need to teach literacy to the benefit of our students, the natural next step was finding time and building collaboration. Cami knew the science, and we understood the literacy. Together, we needed to join our expertise.

For the purposes of developing students to think like scientists—to meet the learning standards of the science curriculum—we wanted to collaborate with Cami to create literacy strategies that were aligned with the outcomes in her science classroom. We knew her planning time was limited, so we wanted to make sure that our support in the area of literacy was thoughtfully integrated with the learning outcomes of her science units.

Collaboration plays a crucial role in the success of any school dedicated to building effective teams in a PLC culture. When experts collaborate, innovative ideas emerge in ways that support student learning and generate positive results. When collaborative time is used wisely—when the action steps of a team are clearly designed, intentional, and focused—it is possible to make great progress in student learning. Across North America, schools are making a commitment to this core principle, they are tackling long-standing concerns in education by bringing together teacher teams to make stronger curricular and instructional choices, and they are getting better and better at making use of assessment practices that support the formative development of all students.

This chapter helps you identify how to initiate collaboration by applying PLC fundamentals and build teacher teams within your school to support meaningful collaboration that leads to student growth and reflective teaching practices. We offer guidance for leaders and examine how to approach meeting logistics before delving into the details of the work teams carry out in collaborative meetings, including analyzing standards and setting goals, identifying students' existing literacy skills and needs, and finding connections between the content area and literacy skills.

Collaboration Within a PLC

PLCs are a pivotal force for progress in schools, as they are all focused on three big ideas: (1) a focus on learning, (2) a collaborative culture, and (3) a results orientation (DuFour et al., 2016). Within our literacy work with science teachers, we kept these three big ideas at the core of our work, and we directed our commitment to literacy in science by continuously addressing the four critical questions of a PLC (DuFour et al., 2016).

1. What do we want students to learn?
2. How will we know when they have learned it?
3. What do we do if they haven't learned it yet?
4. How do we extend learning for those who are already proficient?

We recognize that teams are configured in varying ways depending on the school. For instance, you might have curriculum teams, grade-level teams, content-focused teams, or teams that are singletons (teams of one). No matter how your teams are currently structured, when working toward integrating literacy-based strategies, we hope your teams will begin collaborating with a literacy expert in your building—an expert who might be able to provide insight into varying and supportive literacy strategies that can be integrated into your instructional practices. But as we wrote in the introduction to this book, if your building does not have a literacy expert, we designed this book also to function as a thoughtful, collaborative substitute.

While the size and scope of work in a PLC can differ greatly from one PLC to the next, the initial focus of teams often starts with a specific, discrete task that later evolves into more layered tasks and discussion topics. By simple definition, a team is a group of professionals working interdependently to achieve a common goal. PLC architect Richard DuFour (2004) notes successful teams:

- Have common time for collaboration on a regular basis
- Build buy-in toward a discrete and overarching common goal
- Build a sense of community
- Engage in long-term work that continues from year to year
- Grow—but do not completely change—membership each year

In addition to these characteristics, in our experience, it is helpful to have a team leader or point person who creates agendas and monitors discussion (this can be a rotating role); it is also important to encourage open, honest, discussion-based dialogue in order to respect and include all ideas concerning student learning, and we recommend that the core membership find opportunities to reach out and include other colleagues. Remember, collaborative team meetings work better when they are focused on actionable items that will serve to extend the professional learning of the teachers. The following sections go into more detail on the configuration of teams, the role of leaders within a team, and the logistics for team meetings.

Team Configuration

At this point, we want to make a few explicit recommendations for configuring teams effectively when working on literacy-based strategies. There are a number of different approaches to establishing a team, and in many schools, resources vary—for instance, at many schools, there might not be literacy experts trained in the value and capacity of literacy strategies. Sometimes we need to think of teams differently if we are thinking about how to make changes happen. Here are four considerations when building teams focused on literacy.

1. Ideally, incorporate a literacy expert on a curriculum or a content-focused team to serve as an informed thought partner. An individual with literacy expertise can help support instructional changes with a greater variety of approaches and can also help in selecting strategies that are better aligned to the course outcomes. For purposes of the science classroom, the literacy expert can help to adapt strategies that are focused on critical-thinking areas, such as understanding how to read data, synthesizing information, or drawing conclusions.

2. Some teams work horizontally, meaning they work within their specific grade level or within a particular content-based course with many sections and many teachers. Some teams work vertically, where they meet with teachers from different grade levels to ensure curricula are interconnected and working to build learning year to year. When considering literacy-based strategies, consider how your teams are constructed and the purposes of your teams' goals for integrating literacy-based strategies. Why and how should the strategies be applied to support students throughout a particular grade level or within a content-based course that many teachers instruct? When and why should a team use a strategy? If working vertically, how and why might a team teach and reteach strategies year after year? How do these strategies become habits over the sequence of courses and throughout grade levels?
3. Some teams will not have a literacy expert. In these cases, consider what other resources can serve as a strong, literacy-informed, and reflective collaborative partner. Consider the expertise that might come from the ELA department. Is there a reading or writing teacher available to come to your team meetings? If not in your building, is there a literacy expert in a building within your district—an expert in an elementary, middle, or high school building who might be able to help? Is there an expert in the special education department with a background in the area of literacy? Seek to use the resources available to you.
4. Consider how your team might make collaborative use of this book in more directed ways. Or, if you are a singleton team, how might this book be a resource to help you reflect on your current teaching practices and help you to grow in your own learning? To improve, teams often need to seek outside resources that do not currently exist within the schools they are working in. In these cases, widen your definition of a team and consider the larger community of educators who are willing and ready to help other teachers to learn. Reach out to professional organizations or attend conferences that can help connect your team to discussions that will influence positive changes.

Leaders' Roles

Collaboration is at the crux of literacy work and is truly an essential component of professional growth. While this may seem obvious, achieving authentic collaboration can be a challenge for a number of reasons. As discussed previously, we commonly hear "there isn't time" as a primary roadblock. Consequently, the first step in the collaboration process takes place between literacy leaders and administration.

Your past experiences with using immediate strategies to address students' most critical literacy needs provide an opportunity to approach your administrators and create a long-term plan for schoolwide literacy. The following chapters provide information and examples of how to begin this process, but here are some starting points to consider when communicating with administration and when engaging with the following chapters in this book.

- Identify the immediate problem and show the evidence of the problem with data.
- Identify your ultimate goal for the students in your science classroom. Know what you want to change.
- Provide a list of strategies you have tried in your classroom setting. Identify the ways you and your team have worked to build change.
- Provide suggestions for resources or ideas that can help you and your team accomplish your goals and get your students to a more successful place. Collaborate with your administrators over how these goals can be accomplished with the right action steps and the right supports in place.

Teachers must be committed to the challenging work of moving students' literacy competencies in the right direction, and leadership must support and respect this collaboration. Responding to your school's literacy needs cannot occur through an occasional meeting or a purchased program—everyone must be in on the collaboration and be open to their own education and professional growth.

Having said that, we want to offer a few tips for literacy leaders engaged in this work. If you are the literacy leader in your school, consider yourself a host to other teaching teams. You want this team collaboration with you to be as positive as possible so everyone can be an effective participant. Often as a leader in literacy, you will need to serve as the glue that holds the pieces together. You will need to develop the agendas and send out the invites and the reminders. Your personal goal is to keep this team moving forward and committed to literacy-based strategies. Be authentic: you are not the person with all the answers—you really only have half the answer. The key to collaborating with science experts around literacy is to effectively connect the literacy strategies with the content. The credit for the good work that comes from the focused collaborative team goes to the committee as a whole, not the literacy leader.

What additional questions or thoughts could you consider when approaching administrators about creating a schoolwide literacy program that can specifically assist in the science classroom?

Collaborative Meeting Logistics

Based on our experiences, there are four simple logistical action steps you can take that will help ensure fluid and timely collaboration.

1. **Work creatively and collaboratively to carve out a common regular time and space for your team to meet:** From our experiences, carving out common time is the first step to developing a high-functioning team in a PLC. While a progressive education model allows for consistent and regular professional collaboration during the school day, many schools do not have a schedule that reflects this due to the complex nature of the secondary school day. And while a regular weekly team block where students start school late or leave school early is often the ideal way to ensure teacher teams are able to meet in focused ways, we also recognize that many school districts are still working toward supporting structures that support PLC cultures. If your school has not yet set up a structured time for teacher teams to meet, we suggest planning ahead to ensure team meetings can be supported in ways that allow for collaboration about literacy. If you and your teacher colleagues are dedicated to literacy in the science classroom, you might need to work with an administrator who will help support the collaborative time necessary to innovate positive changes. Ask for the time you will need to collaborate and to innovate literacy strategies in your classrooms. For example, you might ask for release periods or release days throughout the year so you can accomplish your team's goals.
2. **Create a regular meeting schedule, and make sure everyone knows the plan:** Collaborative teams must ensure they dedicate their meeting time to fostering the commitments of a continuously developing PLC.

Early on, establishing norms that set focused, actionable goals helps teams achieve their purpose and helps to establish commitments. We suggest using SMART (specific and strategic, measurable, attainable, results-oriented, and time-bound) goals as a good guiding tool (Conzemius & O'Neill, 2014). By setting up SMART goals, your team will be more likely to stay focused, learning, and driven to succeed. As literacy coaches, we know that time is precious, and when our science teachers meet with us, we know that our collaboration needs to have purposeful, specific outcomes. In building SMART goals toward literacy, we encourage taking the time to create an action-driven schedule that is paced, practical, and respectful of everyone's many different commitments. What is realistic depends on your structure and your team's purpose. From our personal experiences, we think that meeting less than every other week often means that team members will be unable to prioritize making changes in literacy. Team meetings that are too infrequent mainly consist of recapping what happened at the last meeting and miss the mark on cultivating productive collaboration that leads to changes to literacy practices in the classroom. To ensure all team members are aware of upcoming meetings, create electronic calendar invites, send out a paper copy of dates and plans that members can post at their desks, and send reminders. Always attach your agenda to encourage thoughtful preparation prior to the meeting.

3. **Identify and use a consistent meeting location:** We also encourage finding a consistent space for your team to meet. It is counterproductive when people are always searching for a changing location. Ideally, this space will be free of other distractions, comfortable, and well equipped for the work you will be doing (for example, have access to a whiteboard, projector, laptop, or computer).
4. **Create digital files that capture agendas and notes:** In addition to ensuring equipment is available in your meeting space, from the beginning, create an ongoing digital hub for agendas and notes—use tools like Slack, Google Docs with Google Drive, or Microsoft OneNote with Microsoft OneDrive. Ideally, you want something that will allow all members to contribute independently and ensure full online access to your materials outside of your collaboration time.

thinking BREAK

- Do you see any natural opportunities within your schedule for team time?
- Are there any predictable patterns you notice in your department or school's master schedule that would allow for meeting time?

Standards Analysis and Goal Setting

As we noted before, teams must build consensus about what they want all students to learn—the first critical question of PLC cultures (DuFour et al., 2016). While your team may have a general overarching literacy goal right out of the gate, it is critical that you dedicate time to the specific development of discrete goals that identify what you want students to learn. Often, it is necessary to do some research, a collective inquiry, in order to establish a goal. Before defining any sort of common goal for student learning, spend time examining your school's science curriculum standards, and assess the current implementation of these standards with a productive and critical eye. Your team must unpack the science content standards and identify the literacy skills they require for mastery. Once you have unpacked them, you will need to set aside the complete content standards momentarily to focus purely on the skills, which serve as the vehicle to transport the learner from novice to skilled on the content mastery continuum.

Investigating current practices and detailing desired outcomes requires your team of colleagues to have thoughtful conversations about the curriculum standards in your science department: what your team prioritizes, the skills your team will focus on, the expectations for student performance, and the criteria your team will use to evaluate and support student learning. This means you will need to unpack the wordy and lengthy verbiage of your science content standards and identify what matters most to your specific course. In other words, your team needs to prioritize your curriculum and identify what your team will emphasize.

Teams that are more focused on their curriculum priorities can better focus on the literacy skills that will support the mastery of those standards. When these priorities for learning are clear, the team can then work on connecting the standards with the literacy strategies that will be most useful. For science teachers, some of those standards will be content-driven—what you want all students to know, understand, and do. Other science standards will be process-driven—how you want all students to learn how to learn science. In your team meetings about literacy, your team will want to configure the use of literacy strategies to support these differing outcomes. For instance, a content standard might focus on literacy skills like recall, synthesis, or drawing a conclusion. The process standards will focus on *how* learning develops.

It is helpful to detail your team's understanding and the department's approach to teaching the standard as well as the evidence of student learning you will be able to collect to demonstrate that students have achieved these skills. Fill out figure 1.1 to discuss with your team the content standards you are prioritizing and the evidence of student learning that you expect students to demonstrate. These are the first two critical questions teams should determine.

Content Standard	**Artifacts**		**Opportunities** ☐ **Power Standard**
	Lesson Plans and Course Materials	**Student-Generated Evidence of Standard**	
Unpacked:			
	Strengths and Weaknesses	**Strengths and Weaknesses**	

Figure 1.1: Content standard–analysis tool.

continued ⟶

Content Standard–Analysis Steps (Use this process with your team to guide collaborative discussions.)

1. **Unpack:** What is the standard really saying? Put the standard in your own words.
2. **Identify artifacts:**
   - *Lesson plans and course materials—*
     - How is the standard being taught?
   - *Student-generated evidence—*
     - How are students demonstrating mastery?
3. **Assess the strengths and weaknesses of these artifacts:** Where is there room for improvement?
4. **Identify power standards:** Which content outcomes are the most essential to your content area?
5. **Identify next steps:** As your team's work continues, identify potential teaching and learning opportunities.

Visit ***go.SolutionTree.com/literacy*** *for a free reproducible version of this figure.*

We urge you not to overlook the steps in the content standard–analysis tool. While it is true that content standards may shift, it is also critical that we continue to evolve curriculum alongside changing standards. While it is very tempting to jump to identifying *power standards* (essential outcomes), make sure that you don't skip the unpacking of these standards first. Curriculum teams that have gone in this direction often fall into the trap of simply overstating core curriculum components—essentially saying, "We do all of this already." This is a misstep, as it will not lead toward additional growth or foster a team mentality that is focused on problem solving.

Another way to think of *content standards* is to frame them as desired student outcomes—the concepts that you want your students to have mastered at various checkpoints throughout the year. Conversely, *process standards* will help your students get to this point—they are essentially the vehicle that helps students get to these learning outcomes. These should be composed of a variety of focused tasks that teachers support via carefully sequenced literacy strategies used during instruction. For example, teaching students to read for details that help them fully understand a scientific process is essential for the students' clear understanding.

Identification of Students' Literacy Skills

In order to dive into disciplinary literacy, your team will need to have insight into your students' literacy-skill strengths and weaknesses. Literacy-based inventories or assessments that might show up on a standardized test like the ACT or on a reading passage from the SAT will help you identify where students are in terms of literacy skills so that you can later determine the necessary strategies to navigate challenging text tasks.

Many schools using the response to intervention (RTI; Buffum, Mattos, & Malone, 2018) or multitiered system of supports (MTSS; National Center on Intensive Intervention, n.d.) frameworks already have literacy benchmark assessments in place, although the data are not always shared among all departments in the school. These assessments typically hold a wealth of information pertaining to student strengths and weaknesses for teachers of all content areas. If your school has a reading specialist on staff, he or she would be a great colleague to approach to learn what data your school has already collected and what other resources might be available.

While working with a reading specialist, your team might want to design your own literacy-based inventories, employing the content area's authentic texts and assessing the literacy skills most pertinent to science. Studying these initial student data will be a key component to your team time and will help you identify which strategies scaffold student success and help you further develop and utilize a variety of assessment types, including formative and summative tools.

A great deal of confidentiality and professionalism is a necessity for a productive collaborative model. Getting to know your students better means looking at and sharing your student data with your team. If you want data to help move students forward, then you must handle them with great care, and you must confront the literacy data you collect.

During team meetings, it is critical that no one makes sweeping statements or generalizations regarding teaching practices based on raw data. When viewing group data in a team discussion on progress, remove names or class cohort information. Start the data discussion by noting this information may reveal things the team has already considered. For example, rarely does a nationally normed assessment show that a star student is one of the most struggling readers. Begin the data conversation with the idea that the data will often confirm what you already know about your students, but they may also shed more light on *why* students are struggling or thriving.

As your colleagues become more comfortable with data reviews, fears in sharing data among the team that might exist will gradually drop away, and the team will become more open to collaboration that is focused on student learning. If you are a leader on the team, start with yourself; don't be afraid to show your own student data. Students have entire academic histories prior to meeting their current roster of teachers. There is no way that one week in Mr. Williams's class dictates all aspects of a nationally normed score.

Although it is critical that teams view data in a supportive manner, make sure that your team's analysis doesn't look like a list of excuses. It is very easy to fall into the pit of "things we can't control" when looking at less-than-optimistic numbers. Instead, look at what you can change; the discussion should focus on how you move students toward a goal, target, or outcome—and ultimately toward graduation, higher education or training, and a career.

Identification of the Content-Literacy Connections

In various content teams, some teachers might say, "There really isn't any reading in my subject area." This is based on a traditional view of the concept of literacy. When we say *literacy*, we mean the act of engaging, knowing, and ultimately being able to navigate new understandings of known and unknown nuances associated with defined content.

Yes, this sounds complex! But really, teaching literacy is as much about breaking down an idea, only to build it back up again by scaffolding and modeling a process—just like teaching a child how to tie a shoe. With this said, words are literally everywhere, and many literacy texts are multimodal. Formulating an inference from a reading is similar to formulating an inference from what someone might say, from a film, or from a scientific demonstration. The literacy strategies we use to understand or to infer are often similar no matter what the modality might be.

After your team has established a solid common baseline knowledge of the concept of literacy, focus on your power standards to identify which literacy tasks are the most innate to your area of study (such as drawing inferences, visualizing, summarizing, following various text structures like cause and effect, and so on). Here is where you will be able to identify your process standards and consequently outline the scaffolds that will support students' literacy-skill development. Familiarizing yourself with the strategies that come later in this book will help you work through the process of finding your content-area literacy connections. Additionally, you will need to take a close look at your texts (solo, or as a team for common texts) to determine if they are appropriate for your readers and tasks. Gather all texts, and assess your collection as a whole, using the tool in figure 1.2 as a guide.

1. Does content match your defined content power standards? If not, what is missing?
2. Are these texts at an appropriate text complexity level for your readers?
3. Are your text tools varied to address the different types of text associated with your discipline?
4. Once you have gathered your collection, what scaffolds will your students need to comprehend the content successfully?
5. What scaffolds will your students need in order to transfer knowledge to new contexts and applications?

Figure 1.2: Review tool to discern whether text tasks match complex disciplinary literacy demands.

Visit ***go.SolutionTree.com/literacy*** *for a free reproducible version of this figure.*

Wrapping Up

If a fully supported and committed team sounds like a far-fetched dream, that's OK. All you need is one colleague to join you in your efforts and to formulate collaboration that leads to positive changes! As you are working collaboratively, be sure to share your journey and findings with your other colleagues. Collaboration isn't always as formal as a designated team time—it often starts with simply sharing what you are working on and trying to accomplish with your students. It involves a lot of give and take, with both students and colleagues, but by conducting a thorough analysis of your standards, texts, and students' skills, you will be well on your way to creating a common understanding of where your students are, and you will be able to better determine how to move them toward disciplinary literacy.

Effective teams work together diligently and value all contributions in their quest to help students succeed. Building a productive team is essential to working collaboratively to understand the concept of science literacy, unpack the science standards, set appropriate goals for students, and develop tasks that foster student growth.

Collaborative Considerations *for* Teams

- Who are colleagues you can approach to begin fostering collaborative teamwork?
- What science literacy skills lend themselves to deeper team examination?
- If you already have a science- or literacy-based team, how is it organized? Is there a PLC culture in your school or district that supports dialogue and all ideas that might lead to improved student outcomes?
- How can you use the resources in this chapter to develop your PLC culture?

CHAPTER 2

Foundational Literacy Triage

Throughout this book, we focus on how to establish a culture of literacy in the science classroom as well as how to establish a literacy focus throughout the school. As educators committed to learning, we all need to recognize the value of literacy, and we all need to celebrate the results when a whole school establishes goals and creates an environment for students to feel safe when learning and attempting literacy strategies. However, this process takes time, and more often than not, we have teachers come to us who need immediate assistance. They have specific students, classes, projects, or assignments they are struggling with, and they need a strategy or way to help the students *now*.

Always keep in mind the array of learners in your classroom. When instructing students in a mainstream classroom, there will certainly be a range of different student learning abilities within each class. Further, students' reading levels may be even more diverse, and among these reading levels, you will have students with special education needs (emotional, learning, or medical), students with 504 plans, and students who are only just learning English. As a science teacher, you will need to meet the individual needs and developing potential of *all* the students in your classes.

We have seen that teachers often begin to notice concerns related to literacy early in the year and look for differentiated instructional ideas to meet student needs. This chapter helps you identify ways to *triage* science literacy immediately and work with the variety of student needs and abilities in your classroom. We cover the basics of RTI practices, differentiated instruction, assessment of text complexity, and several "fix-up" strategies to assist science students when reading a text (Tovani, 2000). We close by examining factors to consider when supporting struggling students and for further advancing students who show proficiency.

Understanding Response to Intervention

RTI, also referred to as MTSS, is a three-tiered systematic process for ensuring the time and support students need in order to learn at high levels. Austin Buffum, Mike Mattos, and Janet Malone (2018) explain, "Tier 1 represents core instruction, Tier 2 represents supplemental interventions, and Tier 3 represents intensive student supports" (p. 2).

Because students' literacy skills are often linked to their learning in other academic areas, thinking about the role of literacy in RTI is an important topic to cover when students need support. We believe that teaching literacy is the responsibility of all teachers, and we think working with literacy interventions is a Tier 1 commitment. As we've noted earlier, often more intensive literacy-based strategies are needed to support students who might struggle—where Tier 2 or Tier 3 interventions might be necessary. Identifying students who might need more intensive support in areas of literacy is an important consideration early in the school year—especially in science courses, where readings can be difficult.

Gathering data about students' reading and writing skills can be very helpful in learning more about the range of abilities in the science classroom. Data found on nationally normed tests, such as MAP and STAR, can provide science teachers with early insights into students' literacy skills. Likewise, teams will also want to consider creating common formative assessments that focus on measuring literacy-based skills important to the science classes they are teaching. Throughout this chapter, we outline a number of literacy-based concerns to pay attention to as students are starting the school year, and we offer suggestions for if and when you and your teams might notice other concerning patterns in student learning.

Differentiating Instruction

A good place to start with meeting students' immediate, varied needs is to consider how you can differentiate instruction. *Differentiated instruction* is instruction that helps students with diverse academic needs and learning styles master the same challenging academic content (Center for Comprehensive School Reform and Improvement, 2007). It also provides students with interrelated activities that are based on student needs for the purpose of ensuring all students come to a similar grasp of a skill or idea (Good, 2006). Differentiating does *not* mean providing separate, unrelated activities for each student. Overall, differentiation supports students with learning differences, helps students retain content and skill, reduces the time students require to absorb information, decreases the need for skill remediation, and allows learners the ability to demonstrate learning in a variety of ways.

There are four planning steps you can use when differentiating classroom work.

1. **Determine the academic content or skills:** The first step in differentiating instruction is knowing and understanding the specific academic content you want to teach. This means knowing what learning targets you want students to accomplish. What is the end result you want for your students?
2. **Gauge students' background knowledge:** Next, you need to have a basic understanding of what the students currently know about the topic or content. At times, the students may have no background knowledge on a topic.
3. **Select suitable instructional methods and materials:** Instructional methods need to bridge what students currently know about the topic and what you want them to learn during the lessons. Knowing what knowledge gaps students have and what information they lack will help you plan to use specific methods and materials that are appropriate for instruction.
4. **Design ways to assess skill mastery:** After delivering instruction in multiple ways, the last step is to assess students' skills for mastery. This assessment can take many different forms. We cover assessing science students in chapter 7 (page 127).

When working with these four steps of differentiation, the process overlaps. Students will all bring different knowledge to the content, work at different paces, and understand the material at different times, and the overlap makes it possible to tailor the process to fit each student's needs at a particular time. In the following sections, we examine each of these steps in greater detail.

Determine the Academic Content or Skills

As described earlier, PLCs focus teams on four critical questions (DuFour et al., 2016). The first thing a team should ask is, "What do we want all students to know, understand, and be able to do?" As a science teacher approaching these questions, determining the academic content or skills (or both) is the first step toward focusing decisions around curriculum, instruction, and assessment choices. We find that aligning literacy-based strategies to support these choices will help teams support students' learning of science. As you and your team work through the strategies in this book, consider how certain strategies are better suited for the academic content and the skills you and your team are focused on teaching students. For instance, if you are working on students' ability to synthesize complex information, your team should work with strategies that help students make accurate and clear summaries. If you want students to draw connections between data and conclusions, your team will want to support students with literacy strategies that are geared toward analysis.

Gauge Students' Background Knowledge

One challenge in teaching science is that it's common for a student's background knowledge of science to be limited, making the teaching of new science particularly difficult. When students' background knowledge is limited, they struggle to make immediate connections or draw quick associations to what might be familiar to them—making it a challenge for them to add to their existing understanding or depth of knowledge.

When a student's background knowledge about a scientific topic is limited, teachers and teaching teams need to recognize that they will need to approach prereading strategies in ways that can draw connections to students' interests and limited understanding of a topic in science that might be newer to them. While this is an added challenge, it can be a rather creative one to solve. In chapter 3 (page 48), we provide some useful strategies dedicated to prereading.

Select Suitable Instructional Methods and Materials

Selecting suitable instructional methods and materials demands that teachers and teams ask a few different questions. For example, when considering literacy skills, selecting suitable materials means aligning reading and writing tasks to the appropriate ability level of your students during their grades 6–12 experiences in science. Pay close attention to the reading level of the materials you are selecting for students—they should be at or slightly above grade level. Students who need more support might need supplemental readings at an easier reading level; students who are more advanced readers might benefit from readings that extend their understanding of the science content.

One benefit of working on a team is that the team of teachers can locate a variety of possible readings to help all students advance in their abilities. Likewise, drawing on the expertise of a team can help inform the differentiated instructional methods a teacher uses to help all students develop their scientific potential.

In considering instructional choices, focus your team on discussions that can help support students in ways that advance their ability to confidently approach simple texts and simple questions as well as complex texts and complex questions. In our work with science teachers, we found that this approach to differentiated instruction helped all students, illustrating that there is no one-size-fits-all instructional strategy.

Design Ways to Assess Skill Mastery

Assessing skill mastery addresses the second critical question of a PLC: How do we know students learned what we wanted them to learn? In our work around literacy, we encourage teachers and teaching teams to focus on assessing skills in formative ways, often, and over time. (We cover assessment strategies in detail in chapter 7, page 127.) When teachers work with formative assessment practices, they can respond more immediately to the learning needs of students and support students more effectively and more quickly. Remember that through your commitment to teaching, assessments are meant to encourage collaborative conversations about the literacy of students in the science classroom—helping to focus discussions on how we can support all students and working to address the third and fourth critical questions: What do we do if students did not learn? What do we do for students who did learn? Assessments are meant to engage teams in developing and extending the potential of all students.

Consider an upcoming reading assignment. How can you differentiate the reading for the multiple reading levels within your classroom setting?

Assessing the Level of Text Complexity

An important factor to consider when seeking immediate assistance with instructing your students is making sure the reading level of the text is appropriate. This information may be available to you at your fingertips. Many times, the publishers of a book provide a Lexile measure or reading-level score, and if a resource doesn't, you can usually determine its level using search tools at the Lexile website (https://bit.ly/2PgXSqw). Equally important is to know the Lexile level of your students. Having this knowledge allows you to appropriately select materials that will challenge students but not frustrate them—materials that are at or slightly above their Lexile level. You can determine individual student Lexile levels through standardized testing measures. (Visit https://bit.ly/2ZmS5EF to learn more about this.)

With knowledge of each resource's Lexile level and the Lexile level for each of your students, the challenge for you and your team is to provide a range of texts on a topic that fits the differentiated needs of your students so that no text is too complex (students will not learn the material needed to bridge the gaps in their learning) or too basic (students will not advance their knowledge during their learning time).

There are many online tools available to help teams find text sets around a scientific topic with a range of different reading levels. Such resources allow teachers to support the reading levels to aid students' clarity of understanding on a topic or to extend the complexity of understanding of students who have a stronger grasp of a topic. The following are great resources that are immediate and easy to search on the internet.

Newsela

Newsela (https://newsela.com) is a large database of stories and articles about current events that are created for classroom use. They are arranged by general themes, such as social studies, arts, health, science, law, money, entertainment, weather, and so on. The articles are student-friendly and often provide different options that can support students at many different reading levels. In other words, you can choose one article and get versions of it at various reading levels. This allows all students to access the text at their level and still be reading, discussing, and receiving instruction on the same topic, making it an ideal resource for differentiated learning.

The website is available in a free version as well as a paid version for more specific readings and questions. It offers a variety of search options for texts based on the topic you select. You can also select the general grade level, text level, reading skills you want to focus on, and specific languages in which you would like the article to be available to students. In addition to allowing you to search for a specific topic, Newsela has text sets available. These sets offer an essential question, supporting questions, student instructions, and extension resources. Each text set includes multiple articles around a general topic.

CommonLit

CommonLit (www.commonlit.org) is another free online tool for school use. It focuses mainly on grades 5–12 reading and writing within different content areas—to be sure, science teachers will find this a strong resource. CommonLit allows you to access texts of high interest that are aligned to standards. Students can annotate CommonLit articles, and many texts include questions or other activities. Users can search for texts categorized into a wide variety of themes (for example, community; justice, freedom, and equality; resilience and success; and social pressure) or text sets (for example, ancient Egypt, nature and conservation, space, and westward expansion). This online tool is great for creating texts at various levels and text sets around a common topic.

Using Fix-Up Strategies

Fix-up strategies are another way to assist science students when reading a text (Tovani, 2000). *Fix-up strategies*—strategies to help students get unstuck during independent reading when they find a text confusing—are strategies that proficient readers automatically use. This is a way to not only model but also teach good literacy habits and strategies to students. Cris Tovani (2000) writes that reluctant readers need more explicit instruction and reminders of these strategies, which include the following (phrased for students).

- **Make a connection:** Find connections with what you are reading to other texts, experiences you have encountered, or the world around you.
- **Make a prediction:** Predict what will come next in a text or make a prediction prior to completing a scientific lab.
- **Stop and think:** When something doesn't make sense, stop and think about what you've just read.
- **Ask yourself a question:** When something doesn't make sense, ask yourself a question about the text. Try to answer the question using what you know or new information you've found in the text.
- **Reflect in writing on what you've read:** Reflecting on your own thoughts about what you read can help to solidify connections to a reading, to identify what might be new, or to stop and formulate an opinion. Reflect on what you might agree with or disagree with, on

what stands out in the reading and why, or on what makes you think differently about the topic and why.

- **Visualize:** Stop and create mental images of the concepts described in the text.
- **Use print conventions:** Examine headings and emphasized words to identify important ideas in the text.
- **Retell what you've read:** To help build the ability to summarize and synthesize, after you read something, stop and retell someone (or yourself) what you read in your own words. Try to capture the key point and key details in a way that makes clear sense to you, the reader.
- **Reread:** If something doesn't make sense, reread for clarity.
- **Notice patterns in text:** Consider whether the reading is establishing a pattern. For instance, does the reading argue both sides of a debatable topic, and are there patterns to how the reading is presenting those viewpoints? Are there patterns in the reading that establish cause-and-effect relationships? Why or how does the pattern help guide the development of the reading?
- **Slow down or speed up:** Adjust your reading to slow down or speed up depending on your level of understanding.

Fix-up strategies are always a good reminder of how to read a text. You can create bookmarks for all of your students, listing these strategies as a quick reference to use when reading. A tangible bookmark in their hands when reading will help remind them to use literacy strategies consistently. When they do so, these strategies will become habits students can apply whenever they struggle with reading comprehension.

When using the fix-up strategies described in this section, we recommend a number of different ways to make sure students are using them effectively.

- Help students to realize that during-reading strategies, like those we present in chapter 4 (page 72), are important *skills* to learn. Developing during-reading skills can help students with their comprehension skills and critical-thinking skills. When first teaching these strategies, help students to understand the value the strategy has for their learning.
- Model how to use the skill. Try not to assume that students will simply use the strategy. Take the time to model *how* they can use the strategy.

For instance, model how to read and ask questions or how to read and retell. Often, students can practice these during-reading strategies with collaborative partners in class. With all science-specific strategies, modeling each of the habits is also good practice, and it provides the students with an example of what they can do and what you expect of them.

- Encourage students to focus on using one or two during-reading strategies at a time. Consider the reading you are giving students, and then align one or two strategies that will be most effective for that particular reading. Ask students to practice that strategy.

You and your team can adapt all of these habits or strategies to meet the specific needs of your students in the immediate classroom setting. For instance, a teacher can adapt the retelling strategy for use in small groups or in partners—where everyone reads and then one student retells the group what they read.

Considerations When Students Struggle

Students will likely continue to struggle and need repetition with these strategies even after you implement differentiation, find appropriate texts for their reading levels, and share fix-up strategies. As with implementing any new strategies or processes, these struggles may persist as the student begins to learn how to read and process a text in ways that help to develop a more attentive level of comprehension and the ability to think while the student is also reading. This takes practice and commitment, which is why we think all teachers should be working with students on these strategies for learning. Some additional questions to consider when you encounter students who are having a hard time accessing the text or using basic fix-up strategies include the following.

- **How might you help students recognize what level of text they are reading or able to access?** One way is to notice the vocabulary level of the text. If the students struggle with too many of the vocabulary words in the text, the reading might be too difficult for them, and they might need support. Science texts are often filled with new vocabulary words, so working on vocabulary development in science readings is often an important commitment to student learning.

- **How might you provide immediate help to students who have trouble reading or understanding specific science vocabulary?** There are many ways to improve vocabulary development. The most important way is by having the students experience the word in differing ways and contexts. Likewise, it is important for students to use the new words they are learning. In the science classroom, consider building *word walls* around your classroom—charts of paper that list important vocabulary words students should be learning to help support their comprehension and understanding of science. Refer to the word walls and ask students to use the words regularly during class time and within their collaborative group work. Remind them that in order to think like a scientist, students need to talk like a scientist.
- **How might you help students stay focused during reading and use the fix-up strategy bookmark you provide?** Have students use the bookmark as a reminder or reference for what students should be thinking about while they read. When establishing a purpose to read, ask students to refer to the bookmark as a way to keep their mind focused on that purpose or on the key questions they should be considering.

Although these strategies will be helpful for students, some students will continue to struggle with the demands of literacy—especially as science texts, concepts, and principles increase in difficulty. When working with students who might continue to struggle, it is important to intervene and anticipate difficulties they might confront.

As a teacher and as a team, work to prepare supports that will help all readers to succeed with comprehending and understanding the literacy tasks expected of them. In setting expectations for reading that are directed and purposeful, teachers can help students achieve at a rate that is individualized through appropriate challenges. A reader who is still developing might need to focus on simpler questions during a first reading and then focus on a more complex question during a second reading. Or, a developing reader might need to be clearly directed on three important points to identify in a reading and be asked to find them in a text (while a more advanced reader might be asked to identify what he or she views to be the most important points in a text). In working with varying reading levels, team collaboration can help with generating new and fresh approaches to the assigned literacy tasks.

Considerations When Students Are Proficient

Just as it is important to support the developing reader in your science classrooms, as teachers, we also want to develop the potential and critical-thinking skills of students who demonstrate proficiency and mastery. When working on literacy-based tasks, consider working with your team to create more complex questions and tasks that will challenge your students to extend their critical thinking and capacity. Your team might have other readings that advance the content knowledge of students, but you might also ask the students to process their thinking about a text in more complex ways. For instance, ask students to respond to higher-order-thinking questions that are focused on building connections or associations between concepts in science, or ask students to grapple with more complex cause-and-effect relationships in science. As a team, remember that your collaborative work should address the developing potentials of *all* students. Often, by focusing on more advanced ways of thinking, teachers and teams will uncover creative and interesting insights into revising their curricula to be more engaging, too.

Wrapping Up

Finding texts at students' reading levels is a powerful way to open up new opportunities for the reading process. If you match this with fix-up strategies and share a bookmark listing these strategies, you can provide students with a successful reading process and help them gain the confidence to continue reading their science text. Continuing to provide opportunities to practice and repeat strategies will help students learn the strategies as ongoing habits. In this way, the strategies become more automatic to their own maturing reading process.

Collaborative Considerations *for* Teams

- What fix-up strategies seem most appropriate for a science text?
- How might you go about adapting the texts you use for different reading and ability levels?
- What are some good science texts your team might want to focus on—texts that might be more challenging for your classes?

CHAPTER 3

Prereading

When we first started collaborating with science teachers, they would bring textbooks and articles to us and simply ask, "Okay, what do we do now? How do we go about getting our students to read the material?" For myriad reasons, students had become increasingly dependent on teachers to simply disseminate the material the day after reading was assigned. Instead of reading to learn, they waited for teachers to tell them what they needed to learn through teacher-led, lecture-based slide presentations (like PowerPoint). Students quickly learned that they didn't have to actually read the assigned material in advance—they could wait until their teacher told them what they needed to know. Moreover, due to testing pressures, curriculum demands, and the need to get through as much material as quickly as possible, teachers felt compelled to keep moving forward instead of holding students accountable for *learning how to learn* from the readings they assigned.

The problem with this is multifaceted. First, when teachers tell students what they need to know, we perpetuate an oral-based culture—a culture where students only learn based on what they hear. While this was a long-standing tradition thousands of years ago, telling students what they need to know does not build them up to be critical thinkers and problem solvers who are capable of learning and questioning on their own. Helping students become literacy experts means teaching them to learn how to learn from what they read.

In working with teachers, our literacy team recognized that we needed to focus on how teachers could hold students accountable for learning how to read scientific texts and how they could learn to use effective engagement strategies that would develop their minds. What was seemingly a simple adjustment—stop simply lecturing on the reading—became a monumental task before us because weaning students from teacher dependency is difficult. But we know that for students to be college-ready, they have to be able to read on their own and manage their own learning. With that in mind, we recognized that we first needed to teach students how to engage in prereading strategies in ways that prepared them to tackle increasingly difficult texts independently.

This chapter explains the need for prereading in science. After discussing the rationale for scientific prereading and the need for collaboration when combining science and literacy instruction, we examine the four types of scientific prereading strategies. We highlight the prereading literacy skills embedded within the Next Generation Science Standards (NGSS Lead States, 2013) and offer concrete strategies we developed through team collaborations with our science teachers. The chapter concludes with considerations for addressing students who continue to struggle and students who reach proficiency.

The Importance of Prereading

One thing that students, and sometimes teachers, forget is that prereading is an essential part of the reading process. In prereading, strategic readers set the stage and prepare their minds to take in and negotiate new information, and it is something students can approach in a number of different ways. On one level, it can be as simple as surveying and previewing a text to gain an understanding of the content—helping the reader to understand the scope of the text and its depth. In other ways, prereading strategies might be more creative in order to engage readers and spark their interests. Still, in other ways, prereading strategies may be more pointed and specific, clearly establishing a purpose to read or identifying what specifically needs to be learned from the reading. In grades 6–12, prereading strategies are important for beginning readers as well as more advanced readers because they help them to frame their commitment to learn from the text—be it a simple text or a complex text. Prereading strategies prepare readers for the reading experience and what they will encounter once they are engaged in the text.

From your teacher preparation classes, you might remember the concept of *schema*, the organizational network in the brain where it takes in information, stores it, and connects it to other ideas (Rumelhart, 1980). It is not beyond reason to expect all teachers to teach their students about schema and make the concept more understandable and relevant to them. Sometimes in our work as literacy coaches, we refer to schema as *brain glue*: if students want information to stick, they have to get in the habit of activating their brain glue.

To reinforce this concept, you can regularly model and share your own reading processes with students throughout the year (for example, when talking about the textbook at the start of the year, speaking about new or relevant articles, or discussing labs). When you regularly model the concepts, then you can expect prereading activities to activate brain glue as students begin to build these reading behaviors and become more proficient science readers.

Prereading strategies help to activate prior knowledge so that new learning can take place, they activate previous experiences to help the reader make connections to new ideas or new information, and they work to prepare the reader in ways that help focus him or her with greater intention and help deepen comprehension and understanding.

Collaboration Around Prereading Activities

Collaboration is a central component of a PLC culture. When experts collaborate, they learn from one another and consider innovative ways to approach student learning. As experts in literacy, we've found that many teachers from all subject areas have some knowledge of basic and intermediary literacy skills like the ideas we discussed in chapter 1 (page 15), but we've also found that many teachers need support in how to adapt and develop ways to make use of literacy strategies that are specifically tailored to their subject area and the difficulty level of the texts. For instance, to get students to be more efficient science readers, reading specialists need to collaborate with science teachers to think about how scientists read so they can adapt basic strategies and skills to fit the literacy needs of science students. They need to share their literacy expertise with science teachers to build smarter, more innovative ways to think about literacy strategies specific to the science classroom. In order to do this, our literacy team focused on how we could collaborate more effectively with our science teachers.

Before we could focus specifically on prereading strategies, we had to consider first the entire literacy-building process. Our first collaborative step was to interview our science teachers to learn more from them about their own reading process. We asked our science teachers questions like, "How would you read this text? What would you have to know to understand the reading? How would you annotate or organize the information?" We wanted to know more about how they approached the reading of science, how they learned from reading about science, and what they did after they finished reading a scientific text. From them, we gathered a number of important insights about how they approached the reading process in ways that could direct our collaboration and decision making about prereading, during-reading, and postreading strategies. From this process, we learned three valuable lessons.

1. We learned that our science teachers often previewed their reading before they read—gathering a sense of what the reading was about and why they would need to read it. This prereading strategy, which we explore throughout the rest of this chapter, gives science readers a motivation to read and helps them gain an overview of the science reading.
2. The science teachers told us that they often read for details, and they often sorted and selected which details were most important to the scientific topic. This during-reading skill is one of the many important skills that are essential to the science classroom. In chapter 4 (page 67), we focus on this skill and how to use it.
3. We learned the science teachers often grouped and sequenced science information in a way that would summarize and synthesize the concepts from the reading. They would often stop and take notes. They made lists of important points, and they organized those lists so that they could easily review the information they gathered. We address this postreading skill in chapter 5 (page 89).

By taking the time to discuss the reading process, we began to understand and monitor some of the reading habits we wanted to teach our science students. In doing this, we began to help our science teachers recognize the value of approaching reading in strategic ways, and we were better able to design meetings to extend these discussions in more specific ways dedicated to the reading process—as it can be broken down into three phases: (1) prereading, (2) during reading, and (3) postreading.

At the conclusion of this first collaborative step, we made it a priority to consider how we could focus on prereading strategies that could help establish a purpose for reading a scientific text and ways teachers might engage students prior to actively reading.

The Four Ps of Science Prereading

In approaching literacy strategies, four prereading strategies are particularly important to consider: (1) *previewing*—scanning a text to gain an overview, (2) *predicting*—considering what the reading will say and what information it might provide, (3) *prior knowledge*—identifying what the reader might already know about a topic, and (4) *purpose*—establishing a focused reason for reading. Previewing, predicting, prior knowledge, and purpose are the four Ps of prereading. Some would argue that these four skills are not just for prereading—they permeate the reading process. In many cases, this is true. Skilled readers are constantly predicting and connecting to prior knowledge; less frequently, previewing and reviewing the purpose occur during the reading process as well. The important thing to remember is that if readers are to navigate texts well, they must learn and use prereading skills in a variety of ways to ready themselves to enter the process of reading more successfully. Here are a few points to consider.

- **Model how to use prereading strategies with students:** Modeling aloud can help students understand how prereading strategies are used intentionally when approaching simple texts or more complex readings.
- **Focus on one or two prereading strategies with students:** While the reading process should be deliberate, no one wants it to become cumbersome. The goal is to use prereading strategies flexibly and effectively.
- **Prereading strategies should become habits that don't take excessive amounts of time:** Although it takes practice to achieve proficiency, prereading strategies are designed to help the reader frame a purpose for reading and a motivation to read. The real work of reading to learn occurs during the next step in the reading process.

As your team of literacy experts works with your team of science teachers to create prereading strategies, use the following as a basic guide to the four Ps of prereading.

1. **Previewing:** Students should scan the reading for various purposes—they could review subtitles, examine images, look for keywords, read captions, or examine graphs they might need to understand. A good preview gives readers an idea of the content they are about to tackle. It may encourage readers to think about what they already know about the topic, it may lead readers to consider what the text is about, and it may help readers understand why they are reading the text and what the author hopes to accomplish. Consider assigning the article "Warning: High Frequency" by Christopher Ketcham (2012). Just as our teachers reported the need to preview in order to ascertain what the article would be about and what they already know or do not know about the subject in our initial team meetings, we want the students to similarly preview and ask themselves what they already know about the topic—frequency—and what they

might infer about the words *warning* and *high* in the title. By thinking for a minute or two about these things, students will be better prepared to read.

2. **Predicting:** Readers should constantly be thinking about what they might read about and learn as they make their way through a new text. Anticipating what is to come next in a text is a good way to both keep readers engaged and allow strategic readers to confirm or deny their thinking, which can lead to better comprehension and retention. When using a scientific text, you might ask students to predict the article's content in general, or more specifically, you might ask them to consider an open-ended question that the text will answer.
3. **Prior knowledge:** David E. Rumelhart (1980) posits that all of our generic knowledge is embedded in our schema. By actively recalling this knowledge, readers are prepared to build on the knowledge they already have and store that new knowledge for later use. More simply, by recalling what they know already, students are better prepared to attack a more complicated reading and later connect it to developing schemas. Also worth noting is that if students have rich background knowledge, they are generally better prepared to handle more complex material to add to that prior knowledge. If students have little prior knowledge, be mindful of the difficulty level of the reading so they are not quickly overwhelmed.
4. **Purpose:** There are two aspects of purpose that are important for readers to understand. First of all, readers should have a clear purpose for why they are reading something. Based on our personal experiences, if a student's reason for reading is "because I have to," he or she is probably not going to get a lot out of a reading. However, if a reader can clearly articulate the purpose for reading, the odds are that he or she will be more engaged and benefit from the experience more deeply. The second part of purpose would be for the reader to predict and anticipate the author's purpose for writing. Sometimes the author's purpose is obvious; for example, knowing that a piece is from the op-ed section of a publication is important for reading comprehension because readers immediately understand its purpose is to convey an opinion. Other times, a reader has to work harder to find the author's purpose by identifying a claim or thesis in the text.

Think back to the last piece you read within your content area. As a science teacher and expert, what did you do before diving into the reading? If you cannot remember, think about a piece of reading you use in your classroom. What would you think about before reading it yourself?

- Do you notice yourself gravitating to any of the four Ps?
- What is most important for a scientist to think about before reading?

NGSS Connections

The Next Generation Science Standards (NGSS Lead States, 2013), the Common Core State Standards (CCSS) for English language arts (National Governors Association Center for Best Practices [NGA] & Council of Chief State School Officers [CCSSO], 2010), and local curricula in science that we have reviewed require close attention to literacy in explicit ways—often requiring students to synthesize information, draw connections, and think critically about what they've comprehended. For science educators, this means not only is content important but teaching literacy skills is also part of the task. Driven by the plethora of information available at the touch of a keypad, students must consume and synthesize information to effectively read and write like scientists.

While there are fewer standards related directly to prereading, research shows that effective prereading skills are a precursor to competent literacy because they prepare relevant schema to connect and engage with the text (Pardede, 2017). Figure 3.1 (page 48) highlights some of the prereading literacy connections embedded in the NGSS. In our examples, we concentrate on the Science and Engineering Practices to illustrate the role literacy plays in science classrooms. For example, NGSS elements such as designing and developing require prior knowledge and the ability to preview a problem or issue to address (note the bolded verbs).

In reviewing the verbs, notice how they might inform how you frame a prereading strategy. For instance, *framing a hypothesis* might obviously relate to the prereading strategy of making predictions. *Based on scientific knowledge* might relate to helping students access their own prior knowledge. *Asking questions* might help students with framing a purpose for reading. Ask students questions like, "What do you want to learn from this text?"

Ask questions that can be investigated within the scope of the classroom, outdoor environment, and museums and other public facilities with available resources and, when appropriate, **frame a hypothesis** [prediction] based on observations and scientific principles. (MS-PS2-3)

Use a model to **predict** the relationships between systems or between components of a system. (HS-PS1-1)

Develop and use a model based on evidence to illustrate the relationships between systems or between components of a system. (HS-PS3-2; HS-PS3-5; HS-LS1-2; HS-LS1-4; HS-LS1-5; HS-LS1-7; HS-LS2-5; HS-ESS1-1; HS-ESS2-1; HS-ESS2-3; HS-ESS2-6)

Design, evaluate, and/or refine a solution to a complex real-world problem, **based on scientific knowledge**, student-generated sources of evidence, prioritized criteria, and tradeoff considerations. (HS-PS3-3; HS-ETS1-2)

Source for standards: NGSS Lead States, 2013.

Figure 3.1: Prereading literacy connections in the NGSS.

As you and your team read and review standards—either the ones you might have adopted or the ones you created locally in your school—consider how the language of those skills helps you to align your instructional practices in literacy. In doing so, the work you do in literacy will connect more coherently with your stated learning goals.

Strategies for Supporting Students in Prereading

When working with teachers one-on-one or in teams, the most logical starting place for improving literacy instruction in the content areas is with prereading. In our practice as literacy coaches working with teachers, we examined past practices teachers have been using to introduce reading in the classroom, we examined how they might personally use prereading strategies when they read, and we explored other prereading options that might help to support their work with students in new, more innovative ways. For many science teachers, this is a powerful moment because it's the moment when they recognize the difference between what they do in the classroom, what they do personally as readers, and what they *could* ask students to do when reading for class.

Consider this: you know how you read, and you're an expert at it. But it's easy to forget that you have to teach students how experts read as well. Therefore, one of the most powerful adjustments you can make as a content-area teacher is to begin learning about and creating prereading activities that guide students to become strategic prereaders. An array of different teachers and teacher teams created the following strategies and lessons for classroom use. Each of them combines the knowledge and experience of both literacy experts and content-area teachers. Each strategy also comes with an explanation and provides adaptations that allow for differentiation for students learning English, students who qualify for special education, and students who may already be proficient. We hope the strategies spark ideas for you and your team to use and adapt for your own classrooms.

Frontloading Vocabulary With the Frayer Model

When we meet with science teachers, they often tell us that students struggle with reading scientific texts because the vocabulary is too challenging. Students may struggle to understand scientific concepts or definitions, and they lack the ability to break down a vocabulary word into different parts. This makes understanding and interacting with the vocabulary of the text prior to reading an essential part of the prereading process.

The Frayer model (Buehl, 2017) is a simple, adaptable strategy designed to help students demonstrate their understanding of complex scientific vocabulary in a graphic organizer. This helps the students to adjust or better understand the concepts in isolation and then work them into the context of the reading.

How to Use

The Frayer model graphic organizer identifies a vocabulary word and then asks students to address four parts to aid them in comprehending that word—(1) what it is (a definition), (2) what it is not, (3) characteristics, and (4) examples and images. As we show in figure 3.2 (page 50), it is adaptable to fit a variety of reading objectives (see page 146 for a blank reproducible version of this tool). The graphic organizer format allows readers to deconstruct a vocabulary concept into smaller parts in order to gain a larger understanding of the word or scientific concept. By breaking the word down into different parts, students can draw connections between different concepts and identify gaps in their understanding. It is often successful in a jigsaw manner, where students are allowed to take on different words and then share their work in small groups.

College Prep Physics

Prereading: Projectiles

Name: Amal Anbender

Definition (what it is):	**Word:**	**Characteristics:**
The force that attracts an object to the earth	*Gravity*	There is an acceleration due to gravity. Weight is a measure of the pull of gravity.
Formulas:		**What it is not:**
$g = 9.8\ m/s^2$ $f_g = m \times g$		It is not the same on other planets. It does not impact the horizontal motion of an object.

Definition:	**Word:**	**Characteristics (what it is):**
	Inertia	
Formulas:		**What it is not:**

Definition:	**Word:**	**Characteristics (what it is):**
	Net force	
Formulas:		**What it is not:**

Definition:	**Word:**	**Characteristics (what it is):**
	Acceleration	
Formulas:		**What it is not:**

Figure 3.2: Examples of the Frayer model.

One way science teachers can use the Frayer model for scientific vocabulary is to have students skim through the targets for a unit to find words that they do not know. This also personalizes the learning for each individual. In our practice, after the prereading, our team of literacy experts and science teachers developed a few short activities to have students share or check the knowledge they obtained and the words they learned from the activity. Figure 3.3 offers the steps for this process.

Day 1: Activity in Which Students Decide on Their Own Words (The students should complete this independently, after the instructor explains or models how to complete this activity.)

1. Students read targets from the unit of study and model thinking through the targets to focus their understanding of the purpose of the learning.
2. The teacher models using the Frayer model graphic organizer by skimming for and reading through the section on a specific term. As part of this modeling, students should do the following.
    - Read the definition. This information can come from the sources provided for them, the internet, class discussion, and so on.
    - Reflect on their understanding of the term if they understand how it is defined in the text. If one or more students do not understand, have them consider how they might identify other sources to clarify the meaning.
    - Based on the definition, on the Frayer model, fill in a description of what the term is *not*.
    - Skim the reading again, and select characteristics that apply to the selected term.
    - Draw an example using the visuals in the text.
3. The teacher assigns homework based on what he or she feels is the next step in understanding the concepts.

Day 2: Optional Short Postreading Activities to Check Understanding and Address Targets

1. The teacher has students compare graphic organizers for accuracy. (Students can use their specific graphic organizers from completing the activity in class the previous day or from completing it as homework.)
2. Students rank their Frayer model terms by specific criteria. For example, for covalent bonds, they might rank them by the strength of the bond. (This is a collaborative activity with classmates, but the instructor should model it.)
3. The teacher has students compare and contrast various combinations of the terms in their graphic organizer. (This is a collaborative activity with classmates, but the instructor should model it.)

Figure 3.3: Process for applying the Frayer model.

Visit ***go.SolutionTree.com/literacy*** *for a free reproducible version of this figure.*

Adaptations

You can adapt the Frayer model for both students learning English and students who qualify for special education to help them better understand the content. Although the reasons these two student groups may struggle with a reading are different, this strategy helps by creating a web of knowledge around a main idea or specific vocabulary words that could be very challenging for these students to understand. Using the Frayer model helps students remember, access, and clearly identify specific concepts. But for these groups, teachers should adapt the prompts in each box for accessibility; however, the answers and learning for all students should be the same. When you consistently use the strategy in this way, students struggling with scientific terms become more familiar with the process and how to implement it to aid in their understanding.

For students learning English, filling out the Frayer model boxes both in their native language and in English can further assist with bridging gaps and provide a clearer understanding of the concepts. Another adaptation to consider when introducing the strategy is to provide a model for the students and create a matching game between the vocabulary words you want to introduce and their corresponding definitions. Providing students with such a model clearly outlines expectations for their work and allows science students to interact more with the words.

In addition to adaptations for special education students and English learners, at times, some students may exceed their knowledge base and need a more challenging task. You can adapt the Frayer model strategy by changing the tasks within each of the boxes to demand higher levels of thinking. This is why the Frayer model is so adaptable for all students.

Frontloading With Quick Writes

Giving students several minutes to make and write down predictions about a new reading allows them to tap into prior knowledge, making frontloaded quick writes an effective way to support them as they set a purpose for the reading that will follow. Because students bring varying degrees of prior knowledge to a text, this strategy allows them to assess what they already know about a topic before taking on new material, and more importantly, it helps them sort and categorize that information as part of a frontloading activity.

How to Use

Give students a short quick-write prompt in the form of a question or a sentence starter (see figure 3.4 for an example). This prompt should relate to the concepts they are about to learn, and it can include vocabulary within it they will learn with the content or larger concepts about the reading. As the instructor, you can narrow the focus or keep it as broad as you prefer. The purpose is to get students thinking and showing their thinking about what they may know, thereby establishing a sense of purpose around the topic.

For this activity, students have only a short amount of time to respond, perhaps five to ten minutes. It should be just enough time for them to get what they know onto paper as quickly as possible. (Whatever duration you give, encourage students to use the full amount.) At the end of the writing time, students can share out to the entire class what they believe they know, or they can get into small groups to share their writing with peers.

For five minutes, brainstorm and write your thoughts about the following question:

If the entire human population adopted a vegetarian (or vegan) diet, what would happen to the biosphere?

Figure 3.4: Quick-write prompt.

Adaptations

Because student choice is empowering and engaging, we consider it best practice to provide students with options by offering multiple quick-write prompts or sentence starters. This helps all students to have a sense of ownership over their writing, as well as appeal to varying degrees of prior knowledge. At times, students may need you to provide information about vocabulary words that they will read in the text. Providing the students with specific words to use within their prompt will give them a starting place for their writing. If students don't have a place to start, it can become frustrating, and they may not be able to write at all.

Frontloading With Research

Frontloading with research is an effective strategy to help science students explore inquiry questions for a lesson or unit of study without the need to rely on existing background knowledge. Through this strategy, students receive the opportunity to use a variety of resources to expand and build a new base of prior knowledge about a topic before engaging with the primary reading. It's also a perfect opportunity to instill in students valuable critical-thinking skills because, although 21st century students are digital natives and are generally very skilled at tapping into information, you have an obligation to teach them how to sift through and evaluate that information to ensure its validity and reliability. Taking a portion of a lesson to model valuable research skills as a prereading activity is an effective way to teach these skills to students.

How to Use

Provide students with an essential question that is pertinent to a forthcoming reading task, and give them time to explore the question using a variety of resources. These resources could come from materials you provide, such as other readings, diagrams, or charts, or they could come from students' exploring online resources, such as articles, website searches, and so on. Be sure to use discretion when determining how ready students are to seek their own research resources. For example, for a biology unit on health, you might offer the following prompt: *If the entire human population adopted a vegetarian (or vegan) diet, what would happen to the biosphere? Read through the information found at http://vegetarian.procon.org and summarize the most important and compelling ideas.*

Provide students with an opportunity to summarize their findings and then share those findings with peers or use them as a reference during their primary reading activity. In doing so, students are able to build new prior knowledge that will enable them to set a purpose for the reading that is to follow.

Adaptations

Students need to understand what the question is asking before they are able to evaluate sources. Special education students or students learning English may be able to better build on their prior knowledge by working with you to complete the task orally. Doing so can also give you a better understanding of where the students currently are in their learning.

To begin this process, first, state the essential question orally and explain what the question is asking. Second, clarify what some of the terms in the question are. For example, some students may not understand what *vegan* or *biosphere* means. Then, read a background article out loud to the students, conducting a think-aloud activity as you read the article. Then, have them list compelling main ideas or what they felt was interesting from the article.

Activating Background Knowledge: Prereading Review 1

Students who are able to tap into prior knowledge are more likely to have success when building new knowledge (Pardede, 2017). The idea behind this prereading strategy is to help students recall how previous learning can lead them down new paths of understanding. Taking the time to remind students of previously learned concepts is an effective way to demonstrate that learning in the science classroom is interconnected. This strategy is most useful when new learning depends on prior knowledge, something that is often the case in a science curriculum. Additionally, because there are diverse learners in most classrooms, some students may not have previously mastered the learning, and this allows them another opportunity to review and learn. For students who did master the material, it is an effective review strategy. In both cases, the overall rationale is to build bridges between old and new learning.

How to Use

Keep this prereading review simple. Focus only on the previously learned concepts that are essential for new learning. As the teacher, you will need to identify what students need to know and create a handout that allows them to quickly identify, label, or define those ideas. Students may use a textbook or other source material to complete these tasks, but you can also incorporate opportunities for students to demonstrate their understanding in their own words.

There should be a strong connection between the previously learned concepts and the reading that is about to take place. The example in figure 3.5 (page 56) shows how you can ask students to recall previous reading or knowledge in order to activate their schema and prepare for new learning (see page 147 for a blank reproducible version of this tool). In this example, the review of photosynthesis will set students up for success in a reading assignment that they will complete immediately thereafter. Reviewing these terms will clarify important information and lead to better metacognition.

Adaptations

For students learning English and students who qualify for special education, activating prior knowledge is key. Many times, they may forget or find this information extremely confusing when they initially learn it. Providing examples in visual forms (pictures, charts, diagrams, graphs, and so on) can help jog their memory of the information. Also, providing information in various forms of media

Prereading: Recall or skim section 10.1 on the Calvin cycle (photosynthesis) to help you complete the following steps.

1. In the space provided, draw an illustration of photosynthesis that includes the following labels. (Refer to page 238 if you need a refresher.)
    - The two steps of photosynthesis
    - The *reactants* and *products* of the Calvin cycle. (HINT: There are three reactants and three products.)
    - The sources of energy for the Calvin cycle.

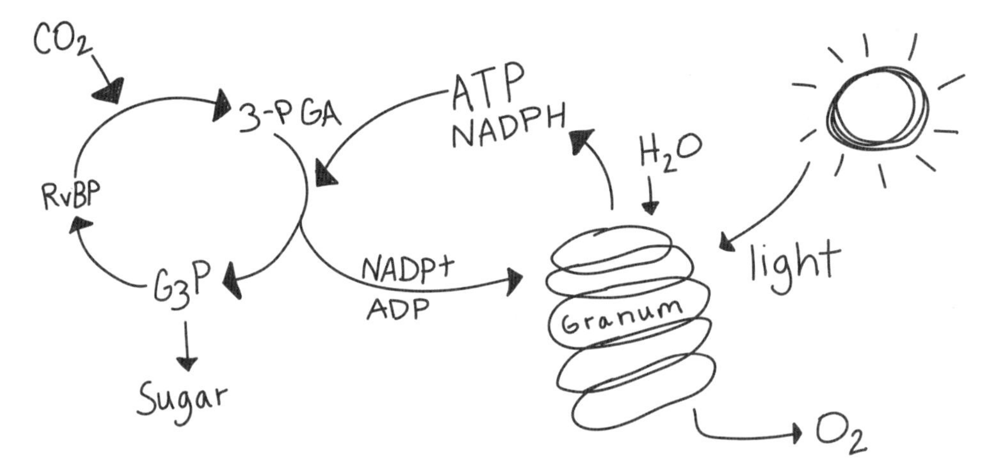

2. As best as you can, define *reduction* in your own words.

    Gaining an electron by atom/molecule

3. Explain how CO_2 entered the plant leaf.

    CO_2 enters through a leaf's stomata.

4. Explain how CO_2 entered the plant cell.

    CO_2 enters the cell through the plasma membrane.

5. Define *carbon fixation* in your own words.

    This is the conversion of H_2O and CO_2 into organic compounds that are easier to use by the organism.

6. Read section 10.3.

Figure 3.5: Recalling previous reading to prepare for new learning.

(such as television clips, movie clips, or YouTube videos) gives students the ability to connect it to what they previously learned from other print and visual resources. In addition, this strategy provides students who are excelling with the material with opportunities to use multiple learned concepts and connect them all together. To facilitate this, provide more than two learned concepts at a time and have the students connect them.

Activating Background Knowledge: Prereading Review 2

Too often, students see reading as something to simply get done. The goal for science teachers should be to encourage students to engage with the idea of reading as a process by explicitly teaching those who have difficulty making connections to prior knowledge how to use text features and thought processes that allow them to tap into previously learned material. To accomplish this, you can provide students with a simple, reproducible graphic organizer that encourages them to preview the reading and connect what they see to prior knowledge so they can make predictions and set a purpose for the reading that follows. The background knowledge they activate should come from previously learned content, material, or experiences. This information will most often come from material students learned earlier in the year. The figures and prompts within the graphic organizer should tap into information the students have previously learned.

How to Use

After you provide the graphic organizer, provide students with time to peruse the text features for their reading, and instruct them to consider how prior learning they've engaged in connects to what they see. Instruct them to note things such as titles, subtitles, images, captions, tables, graphs, and so on that will allow them to make informed predictions about the reading. We recommend adapting the graphic organizer and the directions you provide for using it to meet any unique elements of the reading you are assigning. Doing so allows you to account for unique features that a text may include, such as videos, links, embedded graphic organizers, sidebars, or other elements that may not be a regular part of other texts they have encountered. It's important to highlight these unique features for your students to understand the material better. Ultimately, the idea is to activate all four Ps of the prereading process. Upon completing the organizer, have students share their predictions with their peers before jumping into the reading. Figure 3.6 (page 58) shows a simple graphic organizer tailored for students to look at the

Directions: Skim the assigned reading and note key features, such as the title, headings, images, and captions. Predict the focus (main idea) of the article and what you anticipate learning from the images and captions. Use the provided boxes to list specific features and write your initial thinking and predictions.

Focus:

Feature 1	**Feature 2**
Looking at the bold headlines, I can predict this article is about the relationship between gas pressure and temperature.	The graph tells me that when the temperature goes up, the volume of gas increases.
Feature 3	**Feature 4**
Charles' law has an equation, so I will probably have to use mathematics skills to solve these problems.	The slope of the second graph tells me that, with higher temperature, the gas pressure also goes up.

Figure 3.6: Activating background knowledge strategy.

features in an upcoming reading and connect those to previously learned content. (See page 148 for a blank reproducible version of this tool.)

Adaptations

When students are completing this strategy, they may not be able to make a prediction if they don't understand the reading, or if they haven't gained the necessary prior knowledge. One adaptation you can make to address this issue is teaching students how to turn each of the titles or subtitles from a reading into questions. For example, a diagram titled "The Calvin Cycle" might lead to the creation of the question *What is the Calvin cycle?* Providing them with this strategy gives them a starting point for their thinking and predictions. Students can try to answer the question, thus making a prediction. At the same time, the question provides students with a purpose for reading—to find out more about the Calvin cycle.

Predicting via an Anticipation Guide

By making predictions based on prior knowledge, students are able to set a clear purpose for the reading that follows. The anticipation guide strategy is an approach to prereading (one that continues into the during-reading and postreading phases) that helps students focus on the purpose of a reading. An anticipation guide asks students to read a series of assertions and make predictions about their accuracy. It requires students to think critically about scientific concepts, theories, or ideas before reading about them and then, during the read, look to prove or disprove their predictions. This process helps students focus their reading around evidence and key details that support or refute their predictions, allowing them to take more ownership of the reading.

How to Use

You can start by providing students with a list of statements, asserted as facts. The example in figure 3.7 organizes these statements in a central column. (See page 149 for a blank reproducible version of this tool.) Relying on prior knowledge and existing literacies, students use the leftmost column to indicate if they agree or disagree with the assertions. Providing students with items that are debatable or controversial often leads to the most exciting conversations while the students are filling out the anticipation guide or during a discussion following the prereading activity. Then, during the reading process, students record evidence that either supports or refutes the statement and their prediction. After reading, students use the rightmost column to confirm or revise their initial assessment.

Before viewing: Look at each statement carefully. Put a check in the appropriate column in Before Viewing to indicate whether you agree or disagree with each statement.

During viewing: In the center column, write the evidence from the video that supports or contradicts the statement. Include the time of the video where you found your evidence.

After viewing: Reread the statement and the evidence that supports or contradicts it. Put a check in the appropriate column in After Viewing to indicate whether you now agree or disagree.

Before Viewing		Statement and Evidence	Video Time	After Viewing	
Agree	Disagree			Agree	Disagree
✓		1. An increase in a wave's amplitude means the frequency also increases. Evidence: For sound, an increase in amplitude increases volume, not frequency.	3:17		✓

Figure 3.7: Sample anticipation guide.

continued ⟶

Before Viewing		**Statement and Evidence**		**After Viewing**	
Agree	Disagree			Agree	Disagree
		2. Waves carry energy, but not matter, from place to place. Evidence:			
		3. How does the interaction between matter and energy for a transverse wave occur? Evidence:			
		4. The number of waves that pass a given point in one second is the frequency of the wave, measured in hertz. Evidence:			
		5. Long wavelengths have big frequencies, and short wavelengths have little frequencies. Evidence:			
		6. One wavelength can be measured between points A and B or between points C and E. Evidence:			
		7. Mechanical waves can travel through outer space. Evidence:			

Adaptations

When adapting this anticipation guide for students learning English or students who qualify for special education, color-coding the Agree and Disagree columns can help students quickly identify the columns for indicating agreement or disagreement. Also, it may help students to complete this exercise orally for the first couple of items. Articulating their thoughts out loud can help them better understand confusing ideas or concepts and clarify different thought processes.

Predicting and Confirming Activity (PACA)

The predicting and confirming activity (PACA; McKnight, 2010) is similar in nature to the anticipation guide. This graphic organizer activity supports science students to activate prior knowledge in order to make predictions about new learning they will encounter in a text. In doing so, students set a clear purpose for their reading as they seek out information that confirms or refutes their predictions—important scientific thinking skills. Although this is a prereading strategy, it lends itself to critical thinking throughout the entire reading process.

How to Use

Unlike the anticipation guide, which provides statements for students, for PACA, you encourage students to write their own predictions for the reading based on having previewed it. It may be helpful to preview the reading as a class and then allow students time to write predictions in the left-hand column of the graphic organizer. While reading (after making predictions), students write a plus if their prediction is confirmed, or they can write a minus if their prediction is refuted. In the right-hand column, students record evidence from the text that supports or refutes their predictions. Figure 3.8 (page 62) shows an example graphic organizer. (See page 151 for a blank reproducible version of this tool.)

Adaptations

For students with limited prior knowledge, it can be helpful to model how to develop predictions. This adaptation may look or sound different in multiple classes depending on the teacher. The teacher may want to walk the students through each step and offer extra assistance or prompting, or at other times, the instructor may want the students to complete everything on their own. It is okay for the strategy to be variable. You can also provide a list of ideas that students can reference to write their own predictions. These can include terms or ideas about the content they are learning.

Name: Emre Croal

Text Title: "Getting the Dirt on Carbon" by Susan Gaidos (2009)

Prediction	(+) prediction is confirmed (–) prediction is not confirmed	Support
Carbon is absorbed in the soil.	+	As leaves decompose or rot, carbon is recycled in air and soil.
Carbon feeds plants and trees.	+	Over time, nutrients including carbon break down in the soil and may get reabsorbed by a plant taking in water.
Carbon is good for our environment.	–	Scientists are concerned about the rapid buildup of CO_2 in the environment.

Figure 3.8: Sample completed predicting and confirming activity.

Using a Vocabulary-Concept Memory Game

Vocabulary in isolation is important, but we also want students to make connections between different scientific concepts and the ways in which those concepts work together to support larger understandings. This vocabulary-concept memory-game strategy is ideal to use before a reading near the conclusion of a unit of study, because students have had ample time to work with the different core unit concepts, making them more likely to see how they are interconnected. In this way, students make use of all four of the Ps.

How to Use

Create game cards using important vocabulary and concepts from the reading (see figure 3.9). Content teachers and literacy experts can work collaboratively to decide these terms and concepts to ensure students are prepared to digest the upcoming reading. In pairs or small groups, students take turns flipping over two cards. This strategy differs from the traditional memory game because, instead of looking for matches, the flipper has to build a connection between the two concepts shown on the cards. This can be a prediction, or students can do this after they have looked up the terms and defined them. In addition, students can write their connection between the two vocabulary words or concepts on a whiteboard. Through this process, students work together to link ideas and concepts.

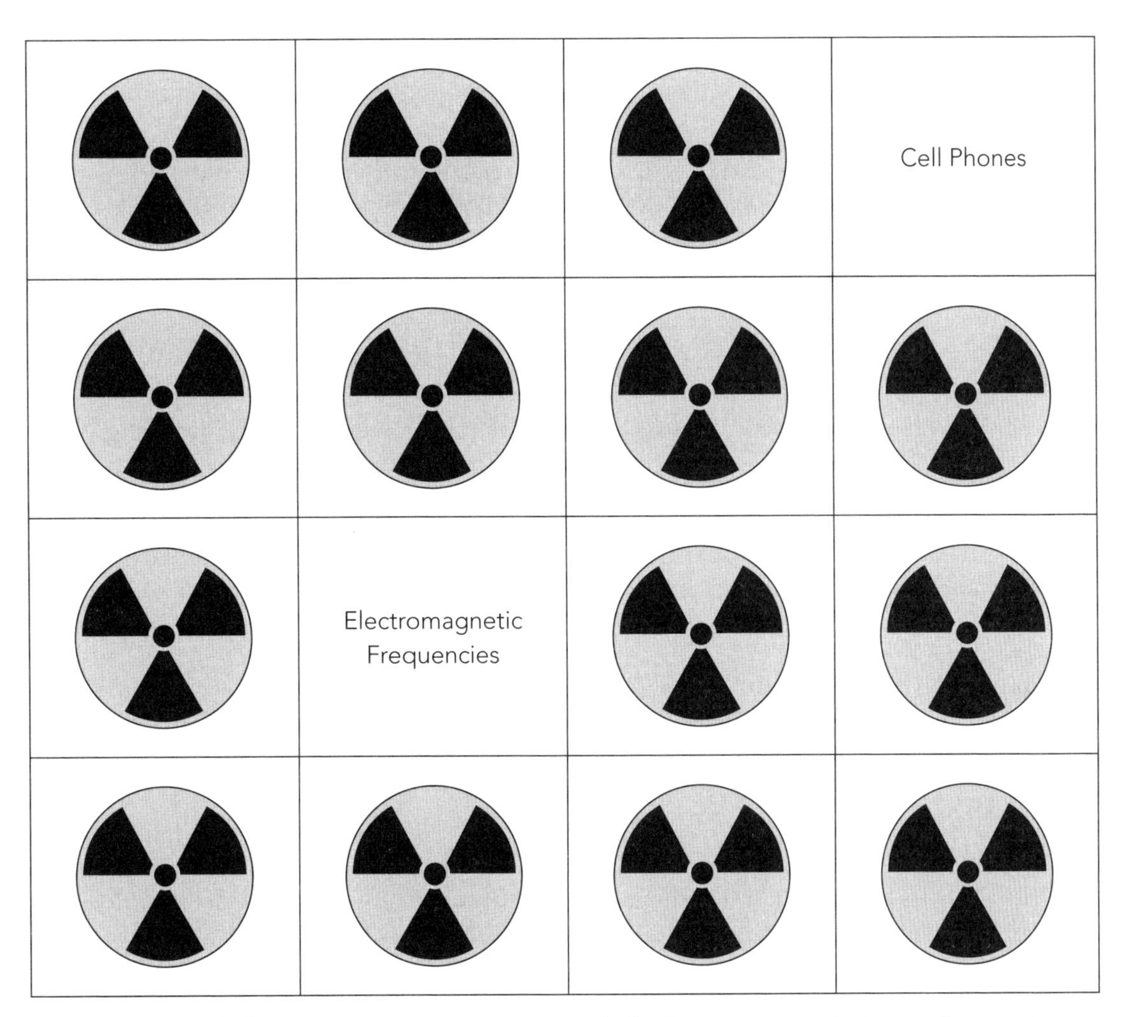

Figure 3.9: Vocabulary-concept memory-game cards for important vocabulary and concepts.

Adaptations

To help students strengthen their vocabulary, add images and phonetic spellings to the cards so students better understand both the concept and the pronunciation of terms. The visuals can prompt or jog the memory of what the word or concept means while incorporating phonetic spellings can be useful for students who understand the concepts but are still learning the language.

Considerations When Students Struggle

The reality is that some students will struggle to use and master prereading strategies and the four Ps of prereading. Such struggles may occur for many different reasons, from not being able to recall prior concepts, to not understanding new concepts, to not being able to make connections between concepts. Here are some additional questions to consider when you encounter students who are having a hard time applying prereading strategies.

- How might you ensure students are regularly stopping and thinking before reading?
- How might you provide students with additional prior or background knowledge they may need before reading? What resources are available in your building to close knowledge gaps?
- How might you encourage students to make predictions before reading?
- How might you help students understand *why* they are reading?

Note that your answers to these questions can be highly variable depending on the nature of your school, what you are working on with your students, and so on.

Review the strategies in this chapter. How might you help students who struggle to make progress?

Considerations When Students Are Proficient

For students who have demonstrated proficiency, PLC teams should collaborate around prereading strategies that help students get to a deeper meaning of a text. Here are some additional ideas for proficient readers in the prereading process.

- Vary groupings so proficient readers are able to work with other proficient readers. Encourage proficient readers to share their prior knowledge and make connections with other proficient readers and their prior knowledge of the text.
- Allow proficient readers to work with nonproficient readers during the prereading process to share their understanding of the concepts.
- Provide extended choice opportunities to proficient readers (in addition to the choices you provide to all other students) that push them to higher thinking levels. In implementing this practice, *do not* provide choice opportunities exclusively to highly proficient readers. The message should never be that only proficient readers are entitled to choice or additional learning opportunities.

Wrapping Up

Using the prereading strategies in this chapter will help students activate their brain glue and be prepared to take in essential content information. You can implement all of these strategies in different classrooms and adapt them for all students to use.

Collaborative Considerations *for* Teams

- Which strategies are best suited for your students? Consider the following.
    - The needs of your students
    - The experiences and knowledge they bring to the reading
    - The experiences and knowledge they need to understand the reading
- Which strategies are best suited for the text?
- Which strategies are best suited for the targeted NGSS?

CHAPTER 4

During Reading

Tom, a biology teacher we were working with who primarily teaches ninth-grade students, noticed a disturbing trend in several of his classes; after assigning reading tasks to his students, many of them were returning to class confused about the reading. Several students were uncertain of what was most important, and others were showing signs that they hadn't completed the reading. Tom knew that the reading and the content were important, so he regularly planned lectures and slideshows to cover the material that students might have missed. However, he noticed that, as the year went on, more and more students had come to rely on his lectures and stopped attempting to do the textbook readings altogether. Tom knew that something had to change.

Make no mistake, the prereading strategies we established in chapter 3 (page 41) set the table for students to engage with their reading, but that work is lost if students don't maintain that engagement into the primary reading. This chapter explores how you can work together with other science teachers as a team to engage students during their reading of all science texts. After looking at the need to build engagement and support active reading, we will review the Next Generation Science Standards (NGSS Lead States, 2013) to highlight embedded during-reading literacy skills in the standards. Next, we offer concrete during-reading strategies we have developed through team collaborations with our science teachers. Finally, the chapter concludes with considerations for addressing students who continue to struggle and those showing high proficiency.

The Need for Engaged, Active Readers

Try to remember the last time you assigned a reading task to your students. Did they respond with enthusiastic high-fives and fist pumps, or were you met with shrugs and a general sense of apathy? For too many educators, the latter experience is the more likely reaction from students when faced with a new reading task. The most compliant students will surely get through the reading, other students may skim through the text to ensure they have a sense of the main idea, and still others rest their heads on their hands, quietly reading line by line.

Imagine your quiet, compliant students reading. Now compare this image with one of yourself at a time when you were fully engrossed in a text. Perhaps you recently read a book or article that you found particularly fascinating. When you become fully engaged while reading, you leave your current surroundings to enter the world of the text. You tune out everything around you and begin to place yourself in the moment. Psychologist Mihaly Csikszentmihalyi's (2009) research reveals that this isn't limited to literary texts; reading an article about a fascinating new scientific finding can just as easily pull us into a state of *flow*, a level of engagement where one becomes completely enveloped in a task or experience. Visualize your own body language when you are completely engaged in a text, and juxtapose that with the image of your students reading with weighted heads on hands. How do you make that image of your students look more like your own, including for those students who wouldn't list science class on a list of things they're passionate about?

Linda Gambrell (2011) argues that students are motivated to read when they have opportunities to be successful with difficult texts. Science teachers have an obligation to provide students with science texts that strike a balance between offering a challenge and not being overwhelming. A sense of purpose is also essential. When students have a clear sense of not only the *what* but also the *why* behind a reading task, they are more likely to be engaged and find meaning in their work. When teachers ask students to make meaning with rigorous texts while also ensuring they have the tools they need to be successful, those students take an increasingly active role in their learning.

In the science discipline, this requires students to engage with a wide variety of text types—charts, graphs, data sets, images, textbooks, or articles to name a few. It also requires taking the time to activate prior knowledge and set a purpose for reading using the strategies in chapter 3. Then, while the reading is actually happening, students need strategies to navigate rigorous scientific texts successfully. Our hope is that the strategies in this chapter will help you and your students find the appropriate balance of rigor and success when engaging in scientific reading activities. In taking these steps and using these strategies, you are already on the right track toward creating a more engaging, active reading experience for your students.

Consider your most recent reading assignment. How did you set it up for your students? Did you give them a specific task or role? Did you simply tell them what pages to read and when the reading was due? What were students expected to *do* while reading?

Collaboration Around During-Reading Activities

Recall in chapter 3 that our literacy team learned three important things (these were *aha* moments for us) from interviewing our science colleagues about their reading habits: (1) the value of prereading to scientific readers, (2) how scientists read for details while sorting and selecting key information, and (3) how scientists frequently summarize and synthesize after reading scientific texts (something we will look at in chapter 5, page 89). For this section, we focus on the second of these things.

Working with Tom and other science teachers helped guide our work toward developing strategies to increase student engagement while also developing essential during-reading skills in our students. Our PLC team went back to the four critical questions of a PLC (DuFour et al., 2016): (1) What do we want students to learn? (2) How do we know if they've learned it? (3) What do we do if they haven't learned it? and (4) What do we do if they have learned it? This chapter is particularly helpful in answering the first, third, and fourth questions.

In each meeting with our science teachers, we discussed matching desired content to a corresponding literacy skill best suited to help students access that content. This work required them to address the first PLC question. In matching content and literacy strategies, it was very helpful that our team consisted of both science content experts and literacy experts, which allowed us to combine our expertise to find the common threads between what students needed to know and the literacy skills they would need to learn it.

As literacy coaches working in a PLC, we had the opportunity to attend a meeting for a team of biology teachers. They were examining formative assessment data, and team members noted that students were struggling to cite relevant evidence from a data table in a lab report—this addressed the first critical question. Examining the assessment data helped us answer the second question. The answers to the first two questions led to our collaboration over the third and fourth questions. In essence, the team recognized the problem but did not know how to teach students to more accurately cite relevant data. Together, we explored different strategies that the team would commit to implementing in future instruction to help struggling students succeed and challenge those who had already demonstrated proficiency.

Replicating this approach with your own team will help you to do the same. But where we had to identify or create instructional strategies to drive during-reading engagement, your team can reference and adapt the strategies we include in this chapter. For singleton teachers, remember that this book is your literacy expert. Use the knowledge and strategies in this chapter to make the connections between student essential literacy skills and science knowledge.

Remember that it is essential for PLC teams to collectively explore student data from formative assessments to identify patterns in learning. What do the data show about students' acquisition of skills and knowledge that help them achieve learning targets? What specific skills need more work or can be extended? Robert Garmston and Bruce Wellman (1998) advocate that teacher teams rely on protocols to help with the process of making these types of team decisions. When teachers have a framework to dialogue around points of inquiry, they are allowed to explore their own assumptions and listen to their colleagues' perspectives and concerns. This allows teams to make data-driven decisions that will have the greatest impact on their collective students' learning. Once your team has worked together to identify the gaps or opportunities for enrichment, it then has the tools it needs to choose one or two learning strategies most likely to address them.

NGSS Connections

The Next Generation Science Standards (NGSS Lead States, 2013) require students not only to become masters of science content but also to become adept at making connections between scientific ideas, disciplines, and systems. They require students to critically analyze data to draw conclusions and make connections. As a science teacher, many of the ways your mind operates when engaging with a scientific text are second nature. For many of your students, though, these do not come naturally. Your role is to provide instruction for your students that makes these thinking habits explicit. The bolded words featured in the NGSS in figure 4.1 illustrate how the standards can connect to literacy instruction of during-reading skills.

Compare the language highlighted in the NGSS with the language found in the anchor standards of the Common Core State Standards (NGA & CCSSO, 2010).

Gather, **read**, and **synthesize** information from multiple appropriate sources and assess the credibility, accuracy, and possible bias of each publication and methods used, and describe how they are supported or not supported by evidence. (MS-PS1-3; MS-LS1-8)

Construct and **interpret** graphical displays of data to identify linear and nonlinear relationships. (MS-PS3-1)

Science disciplines share common rules of **obtaining and evaluating empirical evidence**. (MS-LS2-4)

Analyze displays of data to identify linear and nonlinear relationships. (MS-LS4-3)

Collect data to produce data to serve as the basis for **evidence** to answer scientific questions or test design solutions under a range of conditions. (MS-ESS2-5)

Ask questions to identify and clarify **evidence** of an argument. (MS-ESS3-5)

Ask questions that arise from examining models or a theory to clarify relationships. (HS-LS3-1)

Analyze data using tools, technologies, and/or models (e.g., computational, mathematical) in order to make **valid and reliable scientific claims** or determine an optimal design solution. (HS-PS2-1; HS-ESS2-2; HS-ESS3-5)

Evaluate the claims, evidence, and reasoning behind currently accepted explanations or solutions to determine the merits of arguments. (HS-PS4-3; HS-LS2-6; HS-LS2-8)

Source for standards: NGSS Lead States, 2013.

Figure 4.1: During-reading literacy connections in the NGSS.

Once again, specific words are bolded to emphasize how they align with during-reading skills and the language used in the NGSS.

- **Key Ideas and Details:**
  - *CCRA.R.1*—**Read closely** to determine what the text says **explicitly** and to make **logical inferences** from it; **cite specific textual evidence** when writing or speaking to **support conclusions** drawn from the text.
  - *CCRA.R.2*—**Determine central ideas** or themes of a text and **analyze** their development; **summarize the key supporting details and ideas.**
  - *CCRA.R.3*—**Analyze** how and why individuals, events, or ideas **develop and interact over the course of a text.**

- **Craft and Structure:**
  - *CCRA.R.4*—**Interpret** words and phrases as they are used in a text, including determining technical, connotative, and figurative meanings, and **analyze** how specific word choices shape meaning or tone.
  - *CCRA.R.5*—**Analyze** the **structure** of texts, including how specific sentences, paragraphs, and larger portions of the text (e.g., a section, chapter, scene, or stanza) relate to each other and the whole.
  - *CCRA.R.6*—**Assess** how point of view or purpose shapes the content and style of a text.
- **Integration of Knowledge and Ideas:**
  - *CCRA.R.7*—**Integrate and evaluate content** presented in **diverse media and formats**, including **visually** and **quantitatively**, as well as in words.
  - *CCRA.R.8*—**Delineate and evaluate the argument** and specific **claims** in a text, including the **validity of the reasoning** as well as the **relevance and sufficiency of the evidence**.
  - *CCRA.R.9*—**Analyze** how two or more texts address similar themes or topics in order to build knowledge or to compare the approaches the authors take.

Strategies for Supporting Students During Reading

When science teachers take the time to establish a clear *purpose* for reading, students are already more equipped to understand the *why*. During-reading strategies will help establish the *how*. This section explores during-reading strategies that both build on the prereading strategies in chapter 3 and create a reading experience for students that is active and engaging.

All of the strategies in this chapter are designed to turn students from passive participants into active thinkers. Instead of merely getting through the reading, students will navigate their own way to new understanding. Along the way, these strategies encourage them to question and challenge a text, embrace uncertainties, and engage in a dialogue with the text. As part of teaching and modeling these strategies for students, you can—and should—share your passion for science with your students. Connect that passion for the material to the process for learning it, and you will nurture and grow that same passion within your students. When you explicitly teach students to think like scientists, they are far more likely to find themselves engaged in new learning.

Just like in the previous chapter, science teachers working in collaboration with literacy coaches created or adapted all of the strategies in this chapter. You can use them to support instruction throughout grades 6–12 science curricula. We accompany each strategy with an explanation and provide differentiation options for students learning English and students who qualify for special education as well as make recommendations for students who have shown proficiency with the learning. As with any strategy, it is important to consider ways to adapt these to suit your students and the learning outcomes you need them to achieve. Remember, good literacy strategies apply to simple and complex tasks. Our hope is that these methods become useful tools for your team and your students.

How can the use of prereading strategies help to establish a stronger sense of purpose for *during*-reading tasks?

Text-Dependent Questioning

Students often view reading as a one-way street where the author provides information, and the reader absorbs it. This mentality creates a passive reading environment. Instead, consider how teachers can demonstrate that reading should be a dialogue between the text and the reader. The text-dependent questioning strategy allows readers to become more active, metacognitive, and thoughtful during the act of reading by prompting them to generate their own questions. When science teachers show their students how to read and question a text, they provide a clear model for students to think like scientists. We often tell the teachers we work with that good readers may not have all the answers, but they do have the right questions. For many students, questioning does not come naturally. We owe it to students to teach them how to think and question like scientists.

How to Use

Questioning is a complex skill that has applications across the curriculum. Because questioning skills don't always come naturally for students, they will benefit from seeing a teacher model the ways an expert generates questions while reading. Consider beginning a reading assignment by taking five minutes to model the questioning process for your students using the tools we provide in this section. Be mindful of the nature and complexity of questions that you want your students to generate. Also, when introducing questioning to your students, it may be more impactful to focus on one or two specific question types. We encourage teachers to adapt the strategy to meet the needs of the students.

Figure 4.2 features a text-dependent questioning graphic organizer that students can use to see examples of different types of generic questions and record questions of their own. The tool is organized to categorize different types of questions based on complexity, moving from the most basic questions (*Said what?*) to the most complex questions (*So what?* and *Now what?*). The middle column provides a series of general questions that students can apply to any science-based text. These are designed to inspire students to develop their own text-specific questions that are unique to that text. As students read, they can record their own questions to reflect their thinking while reading. (See page 152 for a reproducible version of this tool.)

In addition to (or instead of) using a larger graphic organizer, consider providing a bookmark with a list of question types or examples. Consider the text you want students to use with this strategy, and choose questions or adapt them from those in figure 4.2. Students will benefit from repeated practice with specific types of questions, but finding a balance between teacher-required or recommended question types and student choice in questioning allows them to establish some control over their approach to learning, focus on developing specific questioning skills, and work at their own level.

Some students may be more equipped to ask higher-level questions than others, and that's okay. Students who have mastered the more literal level of questioning may be ready to challenge themselves with more inferential levels of questioning, while other students may need repeated practice with basic question types. Each student can work at the level where he or she is appropriately challenged.

The questions students generate in the during-reading phase can also play an important role in checking for understanding after reading, meaning you can use them as part of a postreading exercise. Students can also bring the questions they generate to small-group or whole-class discussions, giving them more control in these discussions. We often tell teachers they can learn a great deal about *how* students are thinking and learning based on the questions they are able to generate. For example, if a student is mislabeling a literal question as a higher-order-thinking

Science text-dependent questioning for the text How Roller Coasters Work (Harris & Threewitt, n.d.)		
Question Category	**Questioning-the-Author (Scientist or Experiment) General Questions**	**My Focused Questions**
Said What? What is the scientist or experiment saying?	• What is the scientist or experiment telling you? • What does the scientist or experiment say you need to clarify? • What can you do to clarify what the scientist or experiment says? • What does the scientist or experiment assume you already know?	What is potential energy? What is kinetic energy? What is Newton's first law of motion?
Did What? What did the scientist or experiment do?	• How does the scientist or experiment tell you? • Why is the scientist or experiment telling (or showing) you this fact, statistic, description, example, or visual? • What does the vocabulary reveal about the content or experiment? • How does the scientist or experiment signal what is most important? • How does the scientist construct his or her experiment or develop his or her ideas?	How do roller coasters use potential and kinetic energy? How does gravity impact roller coaster speed?
So What? So what might the scientist or experiment mean?	• What does the scientist or experiment want you to understand? • Why is the scientist or experiment telling you this? • Does the scientist or experiment explain why something is so? • What point is the scientist or experiment making here? • What is the scientist's or experiment's purpose, and what support (evidence or reasoning) does the scientist or experiment present?	How does this connect to other concepts we've learned, like trajectory or inertia?

Figure 4.2: Text-dependent questioning graphic organizer.

continued →

Question Category	Questioning-the-Author (Scientist or Experiment) General Questions	My Focused Questions
Now What? Now, what can you do with your understanding of the scientist or experiment?	• How does this connect or apply to what I know? • How does what the scientist or experiment says influence or change your thinking? • What implications can you draw from what the scientist or experiment has told you?	How might I use this information to determine which roller coasters will be the most fun? How might I use this information to explain why some coasters scare people more than others?

Source: Adapted from Buehl, 2017.

question, it may be an indication that the student has not yet mastered that skill. This strategy makes questioning skills explicit, and it works with almost any reading task.

Adaptations

As previously noted, modeling the strategy is key. This explicit modeling is very important for students learning English and students who qualify for special education, as expressing themselves through questions can be a challenging skill. Many times, students don't understand what they don't know, or they simply lack the ability to express themselves using an unfamiliar language. So, modeling identifies and clearly states the expectations of the strategy. The more examples and modeling students experience, the clearer the awareness of expectations, which leads to fewer misinterpretations of a task due to learning differences or communication barriers. In addition to the modeling, starting with one or two questions in each section narrows the focus for the students, so it may be appropriate to reduce the number of questions on the form. Narrowing the questions makes the task less overwhelming and the outcome you are trying to achieve more explicit.

For students who have previously demonstrated mastery with questioning, encourage them to generate fewer literal questions and generate more higher-order questions, possibly connecting the ideas in one text to previous readings or content.

Note-Taking Charts and Graphic Organizers

In conversations with science teachers, we frequently hear that students struggle with note-taking skills. Even strong readers and writers have difficulty managing all of the key details and important information in a scientific text. Note-taking allows readers to take an active role during the reading process, which leads to greater understanding and engagement for young scientists.

How to Use

You and your team should work together to determine the important content in the reading and then create a graphic chart, web, or other design that allows students to record information in a logical way. By frequently using note-taking structures such as the simple T-chart (figure 4.3, page 78), an expanded information chart (figure 4.4, page 78), and graphic organizers (figures 4.5 and 4.6, pages 79–80), teachers help students build independence and automaticity for note-taking skills.

Most of us have used a T-chart to take notes, and there's a good chance that our students have as well. This simple chart (figure 4.3) relates well to notes that deal with cause and effect, problem and solution, important idea and analysis, and other basic text structures. (See page 154 for a blank reproducible version of this tool.) You can customize each column to fit the information you or your team deems most critical. For the example in figure 4.3, the collaborative team decided to use a T-chart to help students track information while reading *Into the Jungle* (Carroll, 2009).

Sometimes the nature of a text is such that the information in the reading requires more than two columns. Ultimately, the number of columns in any information chart should fit the needs of the content. In figure 4.4, the team has expanded the chart to include four columns because it wants students to draw out and extend their thinking beyond the information in the text (see page 155 for a blank reproducible version of this tool). In this version, the students are asked to read part C of "Seismic Design Principles" (Lorant, 2016) and explain how the design elements of earthquake-resistant structures work and identify where they can be applied.

Some teams and teachers will want to use flowcharts, webs, or different graphic structures to not only help with notes but also help students see relationships and draw conclusions. For these types of graphic organizers, you can creatively construct graphics that fit the needs of the content and help students focus on the important information. Figure 4.5 asks students to not only take notes on chemical structures but ultimately explain the difference between two compounds. In this case, the team created the activity with a specific learning target in mind—to enable students to explain the difference between two compounds. Similarly, we suggest that teams be creative in developing their own graphics and tools to help lead students to collect information with a specific learning target in mind.

Accelerated biography: *Into the Jungle*, chapter 1	
List important events or information from this chapter.	**Why was this important? How does the writer feel about it? Why write about it?**
At 18 years old, Darwin left Edinburgh University without a degree (p. 7).	Darwin's father was desperate for him to be a doctor, but Darwin couldn't handle the gross tasks. Had he remained at EU, he might never have become the explorer we know today.
Darwin's father sends him to Cambridge, where he meets Henslow (pp. 8–9).	Henslow becomes Darwin's teacher and ends up sending him on his famous expedition.
Darwin sets sail on the *Beagle* 12/27/1831 (p. 13).	The *Beagle* is the ship for Darwin's famous five-year expedition which leads to his evolutionary work.
Darwin writes to Henslow (pp. 13–14).	Letters to Henslow, as his mentor, help Darwin think and plan his studies. Henslow also serves as the collector when Darwin sends him species.
Darwin writes his sister (p. 14).	Five years is a long voyage. Darwin probably misses his family. Today, the letters serve as a record of Darwin's journey.

Figure 4.3: Simple T-chart for taking notes.

During the reading of part C in "Seismic Design Principles," complete the following graphic organizer.			
What is the design element?	**Describe or draw the design element.**	**Explain why it works. What does it resist?**	**Where would you use it?**
Bearings	These are stacked coils that separate the foundation from the building.	The separation causes the building's lateral movement to slow down.	Bearings are used between the building and the foundation.

Figure 4.4: Expanded information chart to extend thinking.

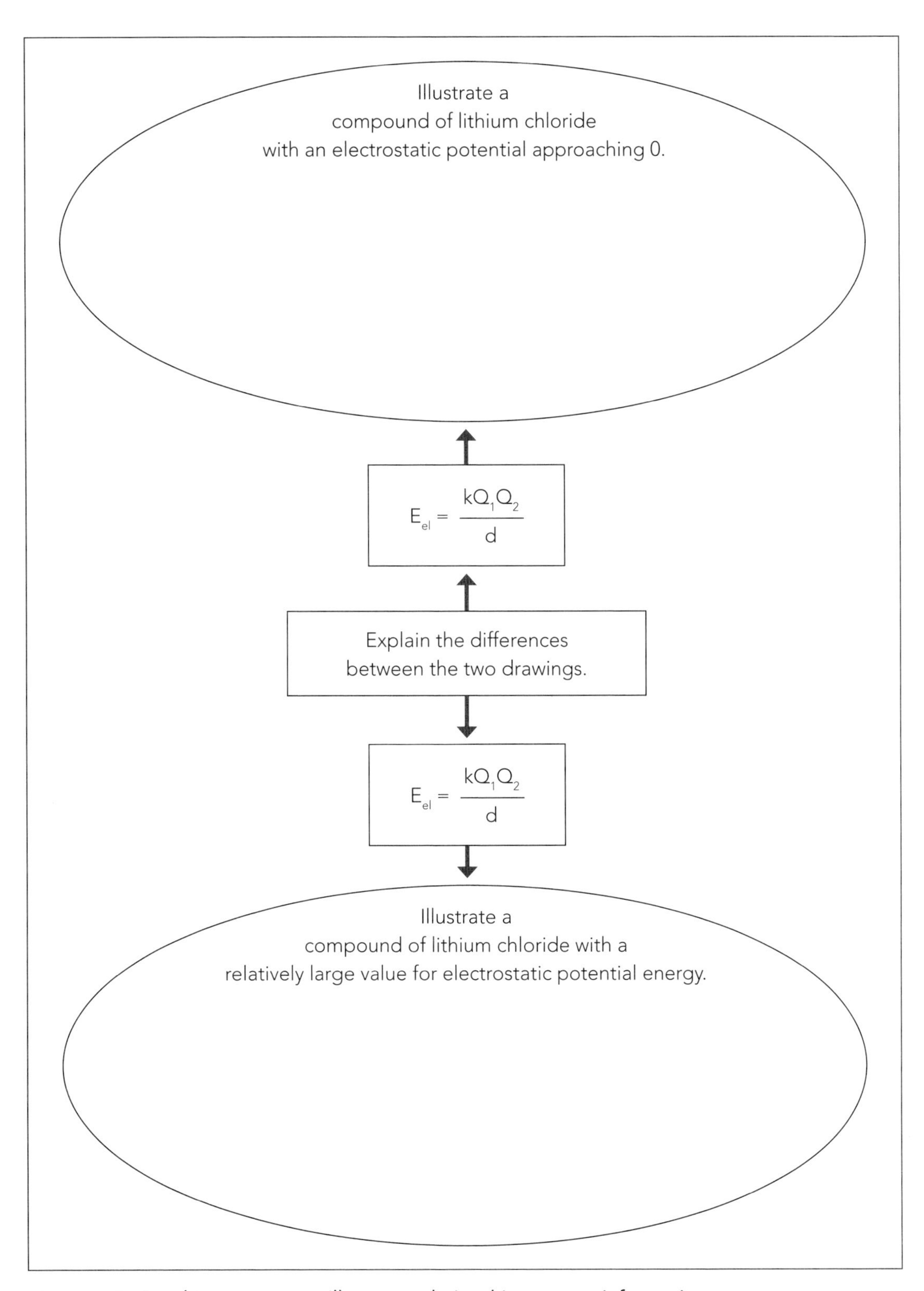

Figure 4.5: Graphic structure to illustrate relationships among information.

One last example of the myriad ways you can tutor students to take notes is by embedding the note-taking chart right into the text when possible. Figure 4.6 illustrates this approach using a short physics reading, called "What Is a Projectile?" (The Physics Classroom, n.d.), which is customized with a teacher-embedded graphic organizer. In this case, our physics teachers used a word-processing program to add the note-taking aid to the web-sourced article, but there are limitless approaches you or your team can take to support student learning in this way.

College Prep Physics Name Christa Stewart

DURING READING—PROJECTILES

As you read "What Is a Projectile?" (go to https://bit.ly/2QvZr0j), use the left column to jot down the main ideas, useful terms, and important concepts. Use the right column to write down your thinking, questions, and any connection to previous units of study.

Recall, we have learned a variety of means to describe the one-dimensional motion of objects. We learned how Newton's laws help to explain the motion (and specifically, the changes in the state of motion) of objects that are either at rest or moving in one direction. Now we will apply both kinematic principles and Newton's laws of motion to understand and explain the motion of objects moving in two dimensions. The most common example of an object that is moving in *two dimensions* is a projectile. Thus, this reading is devoted to understanding the motion of projectiles.

Important terms, concepts, and main ideas	Thinking, questions, and connections to previous units
○ Purpose: To understand the motion of projectiles ○ A projectile is anything that is only influenced by gravity. ○ There are several types.	○ Newton's law: Objects at rest stay at rest, while objects in motion stay in motion unless acted on by a non-zero net force. ○ We've learned about free-fall problems; those are examples of projectiles.

Figure 4.6: Physics reading with an embedded graphic organizer.

Regardless of the note-taking approach students take, begin by modeling how to use any charts or graphic organizers it involves, including what the notes should look like. More important, be sure to show students how the notes will work to their benefit. Will they be useful for an upcoming lab? Will they help students study for an exam? Are there benefits that will reward their close reading and note-taking?

As time progresses, encourage students to be the ones organizing their note structures, including by choosing their preferred note-taking strategy. Encourage students to ask how they can use prereading skills to predict and set a purpose for their reading. Can students turn those predictions and purposes into logically organized charts or graphic structures that help them to organize their thinking while reading? This strategy requires just a quick explanation and provides students with simple, repeatable structures.

Adaptations

Providing structure for students while they are reading helps them understand and know the purpose of their reading. Students learning English and students who qualify for special education may need additional structure at first as they work to organize and put down their thoughts. Simplifying the language or increasing the specificity of column headings on a T-chart or a graphic organizer can help them understand what is important. While some students may be fine with a column that asks them to react to a quote they select from a passage, other students may need more guidance with specific questions, such as, *How did this passage make you feel?* or *What do you think will happen as a result of the passage?*

Eventually, you want students to be able to do this independently, but providing the extra structure at first can help narrow the focus and build the confidence they need to continue the reading and not give up. In addition, teachers can create an added space on the chart or organizer for the students to provide a general opinion or thought about the reading. This opinion or thought may be important for a future conversation or help provide structure for coming to the correct answer.

For students who are proficient in their note-taking skills, allow them more flexibility in the structures of their notes. Students who have not yet mastered the skills more than likely need a structure that is provided for them, but proficient notetakers are better equipped to articulate what is and is not working well for their understanding. That isn't to say a student shouldn't be able to make modifications, but the more proficient reader is more likely to identify where modifications are appropriate.

Text Chunking

Science texts are packed with information. Almost every sentence holds vital information students need to comprehend in order to succeed in the unit of study. For many students, that makes reading an overwhelming task. We've heard students ask, "How do I know what's important if everything is important?" It's a fair question. Chunking the text allows you and your students to break the text into manageable parts, giving students chances to take thinking breaks and compartmentalize what they've read. Not only does chunking the text make reading less overwhelming, but it also helps students slow down their thought process and make meaning from what they've read.

How to Use

Break a larger reading assignment into smaller sections. For each section, students stop and take thinking breaks that allow them to either respond to teacher-generated comprehension questions or generate their own questions and thoughts about what they've just read. This could be as simple as generating a supplementary question guide with numbered paragraphs, or you could provide a note-taking sheet where students take notes or ask questions (perhaps using the text-dependent questioning strategy; see page 73). Students may then use their responses or notes to share their thinking with their peers.

Adaptations

Chunking a text is a natural teaching strategy for any student who is struggling with comprehension. To get even closer to the core aspects of this strategy, take the text and draw actual lines where you would like students to take reading breaks. The visual line makes it very clear to students they should stop and think at this point in the reading. You can add a question at this break, or you can provide a thinking box for the students to respond with their thoughts. This is similar to using read aloud, think aloud strategy breaks (see the following strategy). This concrete structure can become part of the classroom culture, and you can structure all readings in this fashion.

Read Aloud, Think Aloud

The read aloud, think aloud (RATA) strategy is another strategy geared toward promoting metacognitive reading habits. All of us can probably identify students who struggle with reading rigorous scientific texts, yet many of us question the reasons why a student struggles. This strategy encourages readers to verbalize their thinking while reading. Consider how much goes through a reader's mind as he or she constructs meaning or struggles through a text. RATA helps students monitor that thinking and establish mindful habits while reading. By making that reading and thinking transparent and articulating those thoughts, students are able to recognize where they are making sense of the text in addition to where their comprehension is breaking down.

How to Use

Put students in small groups of two to four, and assign each individual a small portion of the text. The smaller the group, the better. Students take turns reading aloud portions of the text, frequently stopping to clearly articulate their thinking while their peers annotate their own text.

At first, students may need help identifying what to share as they read. Encourage them to make connections with prior knowledge, identify areas where they are confused, make connections with bigger ideas in the unit of study, question the text or author, or share any other ideas that run through their minds as they read their assigned portion of the text. As they read, group members listen and annotate their own text. Annotation instructions can vary, so the teacher should be clear about what students are writing and why based on the nature of the text. For example, teachers can encourage students to ask questions, note confusing aspects of the text, make connections, and so on. It's crucial for every student to have a purpose for reading, and annotations help students to be accountable to that purpose. When the reader is finished with the assigned portion, group members have the opportunity to chime in with their thoughts (using their annotations as a guide), perhaps adding new insights or clarifying points of confusion the reader had.

To help sustain engagement, it is best to assign shorter excerpts for each student. A common question we get from students is, "Will this slow me down?" Yes! That's the point. You want to provide students an opportunity to become more aware of their own thinking when navigating a scientific text, and you want them to have an opportunity to hear how others think when reading one as well. For this reason, we encourage you to briefly model the process to demonstrate how a scientist thinks when reading. This allows you to demonstrate the types of thinking you wish to see from your students. There are also a number of postreading strategies that can make an effective follow-up to this strategy (see chapter 5, page 95), but at a minimum, have each group share out its biggest takeaways with the whole class.

We understand that for many science teachers, providing time in class to read may seem unusual or even impractical, but consider the benefits of observing students as they articulate their thinking and insights using this strategy. This is a powerful formative experience that allows you to make meaningful adjustments to instruction, as well as provide students with meaningful feedback that can help them achieve higher-level thinking.

Adaptations

For students learning English or students who have reading or speech difficulties, reading aloud may cause anxiety or nervousness. For this reason, avoid conducting whole-class RATAs, as these can cause even proficient readers to focus more on recognizing words than on constructing meaning (Worthy & Broaddus, 2001). This is why we encourage the use of small groups, which also allow you to organize students into similar ability groups or into groups of deliberately matched peers at different levels. The latter of these provides an opportunity for proficient readers to share expertise and take on a leadership role. Both arrangements can build confidence and raise reading fluency, and they create a more comfortable environment for students when they are working with peers of similar ability or those they can trust to support them.

Also, ensure students who may struggle to read a text have an opportunity to read their section to themselves prior to reading aloud with their group. This first independent read allows students to develop some familiarity with the text, so they aren't so worried about how to say the words. You can also provide students the opportunity to ask their group members how to pronounce some of the more challenging multisyllabic scientific words. Adapting this part of the strategy, along with the actual task of reading aloud, further assists students to succeed.

Embedded Questions and Tasks

Sometimes students need guidance answering important questions or summarizing materials. The fact is, science reading can be dense and difficult. To help students with targeted skills, like summarizing, questioning, and note-taking, it may be helpful to embed the tasks directly into the text so students can begin to see how and where to find important information and how to deal with it logically.

How to Use

When you collaborate with your team and look at student work samples, patterns may emerge that show students are struggling with certain skills, such as summarizing, questioning, and note-taking. To support students and target these skills, you and your team can develop content-focused reading passages that require students to look up words, think creatively, or otherwise attack the targeted skills.

For example, in an earth science unit on carbon, you might assign a reading such as "An Introduction to the Global Carbon Cycle" available as a PDF at the University of New Hampshire (n.d.) website. To create an embedded task from this reading, you would provide students with copies of the text with different kinds of prompts integrated throughout. In this example, at the end of the first section, you might prompt students to create a tweet (280 characters or less) that summarizes the main idea. Then, after the sections on carbon pools and fluxes, you might insert prompts like the following three.

1. Define a carbon pool in your own words.
2. Define a flux in your own words.
3. How do scientists deal with the thousands of participants in the global cycle?

By putting tasks directly into the reading in this way, you prompt students to stop at key points and engage critical-thinking skills to ensure they understand what they've just read.

Adaptations

When using this strategy with students learning English and students who qualify for special education, you may want to embed more questions or breaks throughout the text. This allows for more specific stop-and-think breaks to answer questions or clarify understanding. If there are graphics or pictures within the text, this may be a good starting point. Have the students use the visuals to answer questions and then bridge the visuals with the text. Using both the visuals and text to answer specific questions will help adapt the strategy for readers who struggle with scientific language.

In addition to the adaptations for ELs and students who qualify for special education given in this chapter, what other adaptations can you provide for your specific student population?

Considerations When Students Struggle

Students who struggle with during-reading exercises will benefit from repetition (reteaching) and exposure to multiple reading strategies that engage their higher-level-thinking skills. As with prereading, struggling during reading can occur for a variety of reasons: the text itself may be too high level and create frustration, some students may need additional instruction and practice with a specific strategy, or the vocabulary may be too technical or challenging. Every student is bound to encounter his or her own unique challenges. Here are some additional questions to consider when you encounter students who are having a hard time applying during-reading strategies.

- How might you help students improve the focus of their annotations or notes?
- How might you help students who have vocabulary deficits and difficulties?
- How might you help students with note-taking shorthand in order to speed up their abilities to read and take notes?
- How might you help students stay focused during a reading?

We believe that the strategies in this chapter provide practical instructional approaches for teams working within a PLC to turn to when trying to increase student engagement while reading. At the same time, teachers should solicit feedback (gather data) from the students to find out what is working best and where they find the strategies to be less helpful. This sends the message that the teachers value the student feedback, and it will provide students with a sense of ownership over their learning. Teachers can report that feedback to their collaborative team, and the team can use that information, along with other formative assessment data, to make informed decisions about where to take instruction in the future. The response may be to adapt an existing strategy to a specific case, or it could be to set it aside in favor of a different strategy that may have a greater impact on student learning. It's important to remember that students need practice with strategies—it often takes teachers (in terms of instruction) and students (in terms of learning) multiple repetitions before they can master a strategy.

Review the strategies in this chapter. How might you help students who struggle to make progress?

Considerations When Students Are Proficient

For students who have demonstrated proficiency, teams working within a PLC should collaborate around adopting instructional approaches that are challenging and thought-provoking. In addition to the adaptations in this chapter's strategies, here are a few considerations for further differentiating instruction for proficient readers.

- When varying groupings so that you have proficient readers working with other proficient readers, encourage them to share their thinking with their group while reading. This challenges each student in the group to increase his or her own reading comprehension toolbox.
- When varying groups to match proficient readers with nonproficient readers, encourage proficient readers to share their process for making meaning from texts. This allows them to serve as mentors and leaders to other students in the classroom.
- Develop reading tasks that are appropriately challenging. Many proficient readers are risk-averse. How can teachers push them outside of their comfort zone in ways that will encourage growth? This means not creating more work for proficient readers but ensuring that the time they spend challenges them. Teacher teams should plan for ways that extend learning for any student by building on the reading activity. Having a bank of media texts available is one approach to allow students to continue their learning if they finish reading a text faster than some of their peers.
- Embed choice opportunities that are available to all students. The message should never be that only proficient readers are entitled to choice. This sends a message to all students that they have a say in their learning.

Wrapping Up

Thoughtfully planning during-reading activities allows students to practice comprehension and engagement. If you provide them with ample opportunities to practice and repeat strategies, students will internalize the strategies, ensuring they are strategic readers in future years and courses.

Collaborative Considerations *for* Teams

- Which strategies are best suited for your students? Consider the following.
  - The needs of your students
  - The background and purpose that will engage and focus your students
  - The supports students need to focus on the important content in the reading
- Which strategies are best suited for the text and desired outcomes?
- Which strategies are best suited for the targeted NGSS?

CHAPTER 5

Postreading

It was not long into our collaborative work with science teachers when they asked, "If we are teaching reading, how do we know that the students have understood and obtained the content knowledge that we know will be on assessments?" Obviously, comprehension and access to knowledge are essential to content-area teachers—for students' well-being and for teachers concerned with evaluations based on assessments. Spending valuable class time engaging students in prereading and during-reading strategies that require instructional support and practice requires a leap of faith by content-area teachers that the students will adequately learn content.

The truth is that it would be very easy for teachers to fall back into the habit of lecturing about facts and follow that up with multiple-choice assessments. In such a scenario, students tend to receive good grades and teachers might feel good about their ability to provide students with knowledge, but the problem is that a lot of the knowledge students obtain is short term, and they become dependent on teachers for the information they need instead of learning to access the information from reading and research. Neither of those characteristics creates students who are ready for college, where professors demand students read independently and obtain information to use in class.

This chapter explores how you can guide students toward deeper understanding *after* engaging with a scientific text. We provide a rationale for postreading instruction. We then highlight the NGSS that align with essential literacy skills. We offer concrete postreading strategies that you and your team can adopt to help students make meaning of rigorous scientific texts. The chapter concludes with considerations for addressing students who continue to struggle and those showing high proficiency.

What types of postreading activities do you typically use in your classroom?

- Do you have a pattern of activities that you use with students after they complete a reading?
- Are they always focused on content?
- Do they require any literacy skills?

The Importance of Postreading Instruction

A critical aspect of effectively using postreading strategies in your classroom is to engage students' thinking at higher cognitive levels. To that end, we find that Bloom's taxonomy revised (Anderson & Krathwohl, 2001) is one of the most effective ways to illustrate how students build on their thinking from basic (low-level) to more complex (higher-level) thinking (see figure 5.1).

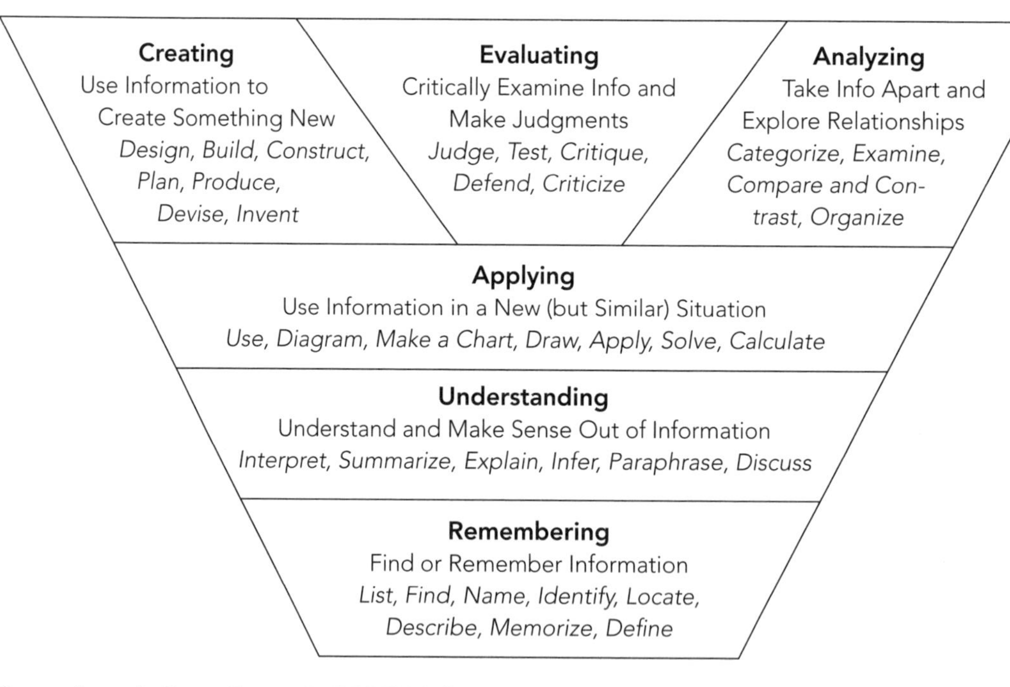

Source: Image by Rawai Inaim, © CC BY-SA (https://creativecommons.org/licenses/by-sa/4.0).

Figure 5.1: Thinking levels in Bloom's taxonomy revised.

Notice that remembering and understanding are the basic building blocks of the taxonomy. Figure 5.1 looks like an inverted triangle because remembering and understanding are the foundation of thinking, but they are basic skills and thus smaller. The levels expand as they move up due to the more complex level of thinking required—represented by a larger level that creates the inverted triangle. A lot of the strategies and the work readers do in prereading and during reading are geared toward gathering information (remembering) and comprehending it (understanding).

The reason the building blocks are important is that without them, none of the higher-order skills are possible. For example, remembering chemical characteristics, components, and structures is a foundational skill for chemistry that students must build from. To understand how chemicals might react to each other or behave in different environments, students must already have memorized chemical properties. Only once students have an understanding of information can they begin to think more critically about it by engaging in skills that work at the apply, analyze, evaluate, and create levels. It is at these higher thinking levels that teachers need students to work during postreading activities designed to engage and challenge students.

Collaboration Around Postreading Activities

In chapters 3 (page 41) and 4 (page 67), we reviewed what we learned during our initial interviews and meetings between our literacy team and our school's science teachers. Recall the first main takeaway centered on the value of prereading to scientific readers. The second revealed how scientists read for details, select key information, and sort information while reading. The last thing we learned from our science teachers was that they tend to synthesize and summarize information after they finish reading sections or whole texts.

Of the four critical questions for a PLC (DuFour et al., 2016), the second question was of particular importance to us for postreading activities—How do we know if they've learned the material? When we sat down with the environmental science team in our PLC, one of the teachers seemed frustrated that her students would often be unable to recall and use important information after reading. Collectively, our teams worked together to figure out specifically what we wanted students to know after reading and what to do if they did or did not understand the content (thereby also addressing the first, third, and fourth critical questions).

Similar to with previous chapters and strategies, we set out with the science team to best match postreading strategies to help students recall and use content material. It is through deploying formative postreading strategies that both teachers and students are able to gain the most understanding of what students do or do not know *before* they take an assessment. In particular, these strategies produce the data necessary for teacher teams to collaborate and make decisions about what and when to remediate and when it is time to move forward to new learning targets and skills. Similarly, these activities help students to self-assess their own level of remembering and understanding to engage their own learn-to-learn skills. As in chapter 4 (page 67), we recommend teams use specific protocols, such as the Garmston and Wellman (1998) procedures, that encourage all team members to dialogue and listen to each other.

NGSS Connections

Looking at the NGSS Science and Engineering Practices for middle school and high school, you will find nouns that convey the need for students to engage in the kinds of critical thinking found at the higher levels of Bloom's taxonomy revised. Words like *evidence*, *theories*, *laws*, *data*, and so on are basic content-based concepts and information that students will have to understand and remember in order to apply, create, evaluate, and analyze. To help you quickly see the literacy connections within these standards, the relevant words appear in bold in figure 5.2.

Analyze and **interpret data** to determine similarities and differences in findings. (MS-PS1-2; MS-LS4-1; MS-ESS1-3; MS-ESS3-2; MS-ETS1-3)

Analyze and **interpret data** to provide evidence for phenomena. (MS-LS2-1; MS-ESS2-3)

Science knowledge is based upon logical and conceptual connections between **evidence** and **explanations**. (MS-PS1-2; MS-PS2-2; MS-PS2-4; MS-PS3-4; MS-PS3-5; MS-PS4-1; MS-LS1-6; MS-LS4-1)

Construct a scientific **explanation** based on valid and reliable **evidence obtained from sources** (including the students' own experiments) and the assumption that **theories** and **laws** that describe the natural world operate today as they did in the past and will continue to do so in the future. (MS-LS1-5; MS-LS1-6; MS-ESS1-4; MS-ESS2-2; MS-ESS3-1; HS-ESS1-2; HS-ESS3-1)

Construct an **explanation** that includes qualitative or quantitative relationships between variables that predict phenomena. (MS-LS2-2; MS-LS2-4; MS-LS4-4)

Apply scientific **ideas** to construct an **explanation** for real-world phenomena, examples, or events. (MS-LS4-2)

Science findings are frequently **revised** and/or reinterpreted based on **new evidence**. (MS-ESS2-3)

Apply scientific **principles** and **evidence** to provide an **explanation** of phenomena and solve design problems, taking into account possible unanticipated effects. (HS-PS1-5)

Construct and **revise** an **explanation** based on valid and reliable **evidence obtained from a variety of sources** (including students' own investigations, models, theories, simulations, peer review) and the assumption that **theories** and **laws** that describe the natural world operate today as they did in the past and will continue to do so in the future. (HS-PS1-2; HS-LS1-1; HS-LS1-6; HS-LS2-3; HS-LS2-7; HS-LS4-2; HS-LS4-4)

Refine a solution to a complex real-world problem, based on scientific **knowledge**, student-generated sources of **evidence**, prioritized criteria, and tradeoff considerations. (HS-PS1-6)

Develop and use a model based on **evidence** to illustrate the relationships between systems or between components of a system. (HS-PS3-2; HS-PS3-5; HS-LS1-2; HS-LS1-4; HS-LS1-5; HS-LS1-7; HS-LS2-5; HS-ESS1-1; HS-ESS2-1; HS-ESS2-3; HS-ESS2-6)

Design, **evaluate**, and/or **refine** a solution to a complex real-world problem, based on scientific **knowledge**, student-generated sources of **evidence**, prioritized criteria, and tradeoff considerations. (HS-PS3-3; HS-ETS1-2)

Evaluate questions that challenge the premise(s) of an argument, the interpretation of a data set, or the suitability of a design. (HS-PS4-2)

Evaluate the validity and reliability of multiple **claims** that appear in scientific and technical texts or media reports, verifying the **data** when possible. (HS-PS4-4)

Evaluate the **evidence** behind currently accepted explanations or solutions to determine the merits of arguments. (HS-LS4-5; HS-ESS1-5)

Scientific argumentation is a mode of logical discourse used to **clarify** the strength of relationships between **ideas** and **evidence** that may result in **revision** of an explanation. (HS-LS2-6; HS-LS2-8)

Make and **defend** a **claim** based on **evidence** about the natural world that reflects scientific **knowledge** and student-generated **evidence**. (HS-LS3-2)

Apply scientific reasoning to link **evidence** to the **claims** to assess the extent to which the **reasoning** and **data** support the explanation or conclusion. (HS-ESS1-6)

Science includes the process of **coordinating patterns** of **evidence** with current theory. (HS-ESS2-3)

Evaluate competing design solutions to a real-world problem based on scientific **ideas** and **principles**, empirical **evidence**, and logical **arguments** regarding relevant factors (e.g., economic, societal, environmental, ethical considerations). (HS-ESS3-2)

Evaluate a solution to a complex real-world problem, based on scientific **knowledge**, student-generated sources of **evidence**, prioritized criteria, and tradeoff considerations. (HS-ETS1-3)

Source for standards: NGSS Lead States, 2013.

Figure 5.2: Postreading literacy connections in the NGSS.

Compare the language highlighted in the NGSS with the language found in the anchor standards of the Common Core State Standards (NGA & CCSSO, 2010). Once again, specific words are bolded to emphasize how they align with postreading skills and the language used in the NGSS.

- **Key Ideas and Details:**
    - *CCRA.R.1*—Read closely to determine what the text says explicitly and to make logical inferences from it; cite specific textual evidence when writing or speaking to **support conclusions** drawn from the text.
    - *CCRA.R.2*—Determine central ideas or themes of a text and **analyze** their development; **summarize the key supporting details and ideas.**
    - *CCRA.R.3*—**Analyze** how and why individuals, events, or ideas develop and interact over the course of a text.
- **Craft and Structure:**
    - *CCRA.R.4*—Interpret words and phrases as they are used in a text, including determining technical, connotative, and figurative meanings, and **analyze** how specific word choices shape meaning or tone.
    - *CCRA.R.5*—**Analyze** the **structure** of texts, including how specific sentences, paragraphs, and larger portions of the text (e.g., a section, chapter, scene, or stanza) relate to each other and the whole.
- **Integration of Knowledge and Ideas:**
    - *CCRA.R.7*—**Integrate and evaluate content** presented in *diverse media and formats*, including *visually* and *quantitatively*, as well as in words.
    - *CCRA.R.8*—**Delineate and evaluate the argument** and specific **claims** in a text, including the **validity of the reasoning** as well as the **relevance and sufficiency of the evidence**.
    - *CCRA.R.9*—**Analyze** how two or more texts address similar themes or topics in order to build knowledge or to compare the approaches the authors take.

Strategies for Supporting Students in Postreading

Before we introduce our approach to postreading strategies, it is important to understand that there is significant overlap between what one might consider a postreading strategy and an assessment. In fact, it is accurate to say that you could use many of the strategies in this chapter for formative or summative assessment purposes depending on the team's learning targets and goals. In chapter 7 (page 127), we discuss our ideas for how strategies like those we list in this section for postreading activities can serve as assessments, but in this section, we focus on what makes a strategy effective as a postreading activity.

Because it is quick and easy, a teacher's go-to postreading strategy is often a multiple-choice assessment. However, the problem with multiple-choice assessments is they generally "fail to assess students' productive skills (e.g., writing or speaking) and to prepare students for the real world" (Abosalem, 2016). Multiple-choice-style questions tend to make students passive participants in theoretical information and often don't assess the more active, complex real-life skills the NGSS and the skills of literacy intend to promote (Abosalem, 2016).

Another issue with multiple-choice assessments as a postreading strategy is that they do not really reflect the nature of information and how we use it today. Students no longer have to memorize everything, because information is available at the touch of a keyboard or through a quick oral query of a smartphone. Instead, as new standards in science, English, mathematics, and other content areas show, the world demands that students can *use* information (International Society for Technology in Education, n.d.). You can see this reflected in the active verbiage we highlight in the NGSS Connections sections throughout this book. Since you want to engage students in using information at higher levels, you also need to quickly determine whether students can obtain such information and use the information. For this reason, postreading really falls into two categories: (1) quick checks to assess comprehension and information gathering and (2) more in-depth checks to allow students to use and engage with the information at challenging levels.

We have accompanied each of the following postreading strategies with an explanation and provided differentiation options for students learning English and students who qualify for special education. As in the previous chapters, strategies include information regarding why and how to use them. We hope these strategies will spark ideas for you and your team to use and adapt in your own classroom.

Five Words

It's easy to take for granted a student's ability to prioritize the information found in a text and then summarize it. There are times in our own classrooms when we are taken aback when we ask students to summarize a passage, only to find they lack the ability to do so. Unsurprisingly, if students do not receive instruction that provides them with the tools to summarize a complex text, many will struggle to prioritize the information they've read. This strategy asks students to summarize the reading by participating in a close-reading exercise. Students will have the opportunity to identify the most important information, share that information with a small group or the whole class, and construct a summary that will include all of the most important information in the text.

How to Use

Before students read, ask them to mark the text anytime they come across a key detail or main idea. We often encourage students to underline or place an asterisk in the margins. This step is essentially an annotation task for students to complete during reading as preparation for summarizing during the postreading activity. When students are finished reading, they must select five words or phrases (teachers can adjust this number as necessary) among those they underlined or marked with an asterisk while reading that represent the *most* important information. This requires them to evaluate all of the information in the text and prioritize what is most critical to understanding. Students can record their individual responses on a form you provide (figure 5.3; see page 156 for a blank reproducible version of this form). Tell students that they should be prepared to defend their word choices for the group work that comes next.

Next, have students work in small groups of three to five students to come to a consensus on a list of five words. Include discussion questions on the response form to guide students in this process. Students can begin by acknowledging the most commonly selected words, but when they have a different word than their peers do, they should defend their selection, and the group will decide whether it is worthy of inclusion on the list. Is that word essential to understanding the text? Is it perhaps encapsulated in a word the group has already chosen? The conversations about word selection are the most valuable part of this exercise, as students conduct a close reading of the text based on their individual selections. Once the group comes to consensus and records its words, students are ready to share with the class.

Each group should display its list, and you can lead a discussion with the whole class, asking questions such as, "What words did most groups agree were essential?

Directions: While reading, underline key words and phrases. After reading, choose the five most important words from the reading, and add them to the Individual Selections column. When instructed, as a small group, discuss the words in the Individual Selections column and come to a consensus on the five most important words that the group agrees on. Add those words to the Group Consensus column.

Individual Selections	Group Consensus
1. Acid	1. Acid
2. Base	2. Base
3. pH	3. Protons
4. Reactions	4. Ions
5. Neutral	5. pH

Discussion Questions

Which words can your group agree on?

We all agree on acid, base, and pH being the most important aspects to understand about the traits of a liquid.

Which words led to disagreements?

Some of us thought our reading emphasized the importance of understanding reactions and what made a liquid neutral, while others thought the chemical processes involving ions and protons were more important.

How did your thinking change as a result of your discussion?

After discussion, I agreed with my group that the concepts of reactions and neutral liquids weren't "big" enough to be critical to this reading. We understand these already, and the point of this activity is to understand what's happening to a liquid that makes it more acidic or more basic.

Figure 5.3: Five words recording sheet.

Are there outliers that need to be explained to the rest of the class? How can this information support a summary of a complex reading?" Of course, there are no "right" answers, but the class will evaluate the selections to determine what is most important. Again, the conversations are the most valuable aspect of this strategy. From here, you can choose to have the class write a more formal summary or leave it at a discussion.

Adaptations

When first introducing this strategy to students learning English and students who qualify for special education, you may simplify the strategy and have them only select three key words or phrases. Starting smaller may make it easier for them to focus on big ideas. Also, providing the students with the first word or set of words gives them an example to follow and use when they are coming to a group consensus. If you give them the structure for their thinking, it helps extend their learning and processing of the information.

For students who have mastered the art of summary, teachers might add on a critical-thinking task that requires students to analyze, question, or evaluate the topic in some way. For example, a teacher might extend an article about recycling's effect on the environment to ask students to question their own recycling habits and the effects those habits have on the environment; or, students might analyze what their school can do to recycle in a more productive way.

Student Self-Questioning Taxonomies

Throughout this book, we have emphasized the value of student-generated questions. We contend that if students know how to ask the right questions, they are on the road toward reading and thinking like scientists. Our colleague Doug Buehl (2017) has written extensively on discipline-specific self-questioning taxonomies. He emphasizes that every discipline requires students to ask questions in its own unique way, and students need to be explicitly taught how to ask different types of questions. Compared to text-dependent questioning (see page 73), this questioning strategy puts increased emphasis on achieving higher levels of critical thinking by asking students to be the ones to build thoughtful questions. It prompts students to reflect on a text they've read and ask questions that they think would be important to a scientist—from what is important to remember to what new knowledge or meaning it creates.

Using a variation of Bloom's taxonomy revised (Anderson & Krathwohl, 2001), this strategy asks students to ensure their questions reflect specific levels of the taxonomy after reading a specific text. The six levels of questioning—(1) remembering, (2) understanding, (3) applying, (4) analyzing, (5) evaluating, and (6) creating—help students identify different layers of comprehension. They also allow you and your students to understand where comprehension breaks down. For example, a question such as, *What is carbon?* is a basic definition question in that it asks about a term or concept that should be remembered. However, if a student labels that question as *analyzing* or *evaluating*, the teacher can intervene and help the student practice developing higher-level questions. In this way, the questions students are able to generate tell teachers all they need to know about what a student does or does not understand.

This ability to formatively assess in postreading has even broader implications for teacher teams working in a PLC. For example, a student who focuses on photosynthesis while reading an article predominantly about recycling may indicate an obvious misunderstanding of the main thrust of a text. Teachers and teams can work together to examine different types of student questions to ensure that the product received meets the learning targets and displays student understanding. If not, the team may have to find ways to intervene and build in more supports and opportunities to practice.

Some of the teachers we have worked with tell us that asking students to write six different types of questions can be overwhelming and time-consuming. We have found that when teachers focus instruction around two to three specific question types, they are able to focus in on specific questioning skills appropriate to the students and task. This approach works best when differentiating instruction. The more proficient students will benefit from working higher in Bloom's taxonomy, while different subgroups will be able to focus on building the skills necessary to reach those higher levels themselves.

In short, if the goal of a science classroom is to train students to read, write, and think like scientists, we believe one of the most important things we can teach our students is to know how to generate meaningful questions about scientific texts. Like most of the strategies in this book, this is not limited to traditional texts. Consider ways students can practice questioning with lab reports, videos, charts, graphs, and other nontraditional texts.

How to Use

Good questioning begins with good modeling. As the instructor, it's important for you to take the pulse of your students to identify where they need focused instruction around different question types. We recommend homing in on two to three question types along the taxonomy at a time. You should determine question types by the task, the text, and the expectations you set for students. If students are working with a complex text and a task they are less familiar with, consider working with lower levels of the taxonomy. If students are familiar with a concept and have had practice with a task, it may be a good opportunity to begin practicing questioning at higher levels of comprehension. Model your process for generating questions for students, and allow them time to practice. We find it's best when teachers model using the text that students will be reading. The teacher can read the beginning of the text and share his or her thinking process when generating questions at each of the thinking levels he or she wants students to target. This makes the process and thinking transparent for the students.

Allow students to work independently or in groups to read a text and generate their own questions in each of the assigned categories. The teachers or student peers should provide feedback on the different questions students generate.

There are many different ways to work with student-generated questions.

- Use student-generated higher-level questions as writing prompts. Students can write paragraph responses to their own questions. Likewise, students can respond to their peers' questions.
- Use student-generated questions in a small-group discussion. Begin at the bottom of the taxonomy to clarify any misunderstandings of what the text says, and work your way up to analytical or big-idea questions that are more implicit in the text.
- Simply collect and check the questions to see how well students are able to understand the various question types. Are they mislabeling a question? Are they unclear about the content? Are students demonstrating they are ready to take on more sophisticated question types? Sometimes the questions themselves tell us all we need to know about our students.

What's most important is that student-generated questions allow you to have a powerful formative tool that puts students in control of their thinking and learning.

Figure 5.4 is an example of how one science teacher uses self-questioning to help students make meaning from a data chart. (See page 157 for a blank reproducible version of this tool.) In this figure, students are asking a variety of questions that help them organize their thinking. Notice how the teacher modifies the question labels from Bloom's taxonomy to language that fits the science classroom. In this instance, the teacher wanted to modify the terminology to match the academic vocabulary of his classroom. *Data* questions encapsulate *remembering* or *understanding* questions, *pattern* questions reflect the *analyzing* questions, *reasoning* questions mirror *evaluating* questions, and *lingering* questions are similar to *creating*

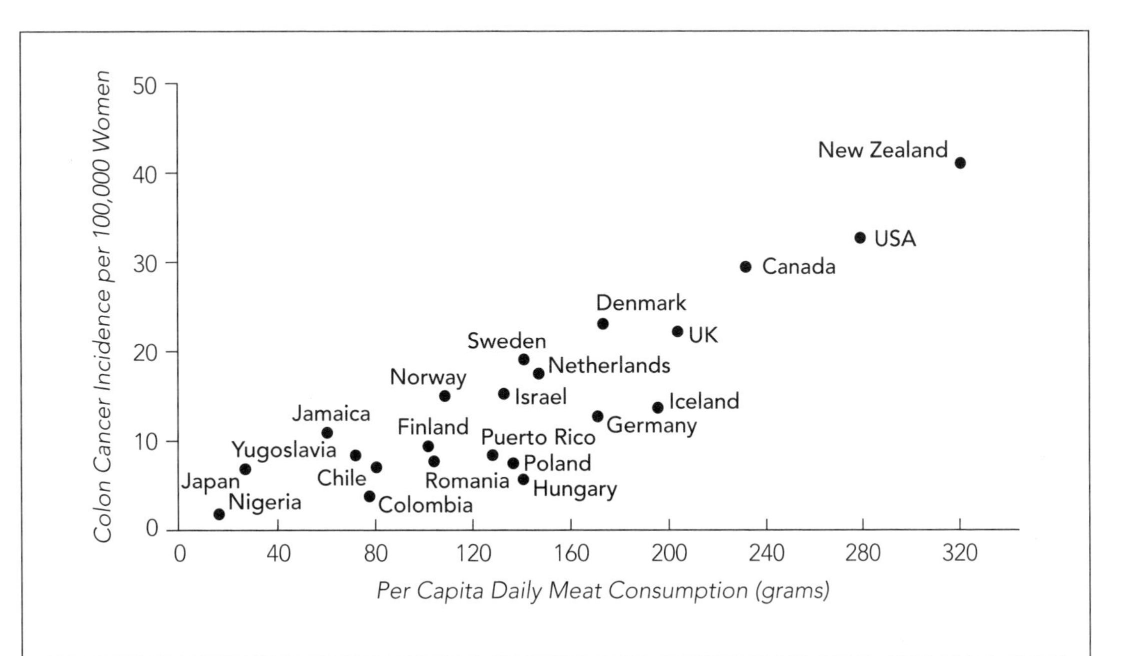

Data Questions	**Pattern Questions**
What data are provided on this chart? ○ What does per capita mean? ○ How much is 100 or 300 grams? ○ What percentage am I dealing with if it is 50/100,000 women?	What questions do you have that deal with patterns? ○ What is the overall trend? ○ How dramatic is the increase?
Reasoning Questions	**Lingering Questions**
How do the patterns (and other aspects) lead to conclusions? ○ What does an increase tell me about the relationship between eating meat and colon cancer? ○ What biological concepts do I know that could support (or possibly go against) these data?	What else are you wondering? Confused by? ○ Why was this graph only done on women? Does gender play a role in colon cancer? ○ What kind of meat were the women consuming? ○ By what mechanism would meat cause an increase in cancer?

Source: © 2017 Thomas Wolf and Brian Wise. Used with permission.

Figure 5.4: Self-questioning tool.

questions because they extend the learning. There is a range of question types that allow students to think at the literal and inferential levels. The questions students generate will provide valuable insights into how they are making meaning from the chart.

Adaptations

This strategy works best when differentiating instruction. The more proficient students will benefit from working at higher thinking levels, while different subgroups will benefit from additional instruction and practice in the lower levels as they work toward higher levels. It is okay to have different students working on different levels of the taxonomy and reading the same text. This gives all of the students an example of others' thinking and questioning and provides a path for students working at lower thinking levels to reach higher levels.

Foldables

Providing students with a kinesthetic activity to apply their learning to helps them better understand, organize, and remember important information (McGlynn & Kozlowski, 2017). A *foldable* is a student-made 3-D graphic organizer that allows students to write information on paper and fold the paper in different ways to organize and reveal the information. Students can add important information such as phrases, specific vocabulary words, or pictures to remember the main points.

How to Use

Students can use foldables as a way to demonstrate their understanding after reading by writing a word or phrase on one flap of the foldable, and then adding explanations of the phrase or the definition of the vocabulary word on the other side of it. Depending on the topic students are learning about, you can have a backside to the foldable with a visual image. This provides various outlets for students' different learning styles. The key to foldables is that they can be as simple or as complicated as the teacher wants them to be. As you can see from the examples in figure 5.5, you can have many different levels of learning within a foldable.

Adaptations

Foldables are adaptable for all classrooms. Providing the phrases or words you want students to focus on is a place to start, particularly for English learners. The students then fill in the explanation or draw a diagram for the phrase or word. For students who may have physical difficulties creating the foldable, providing

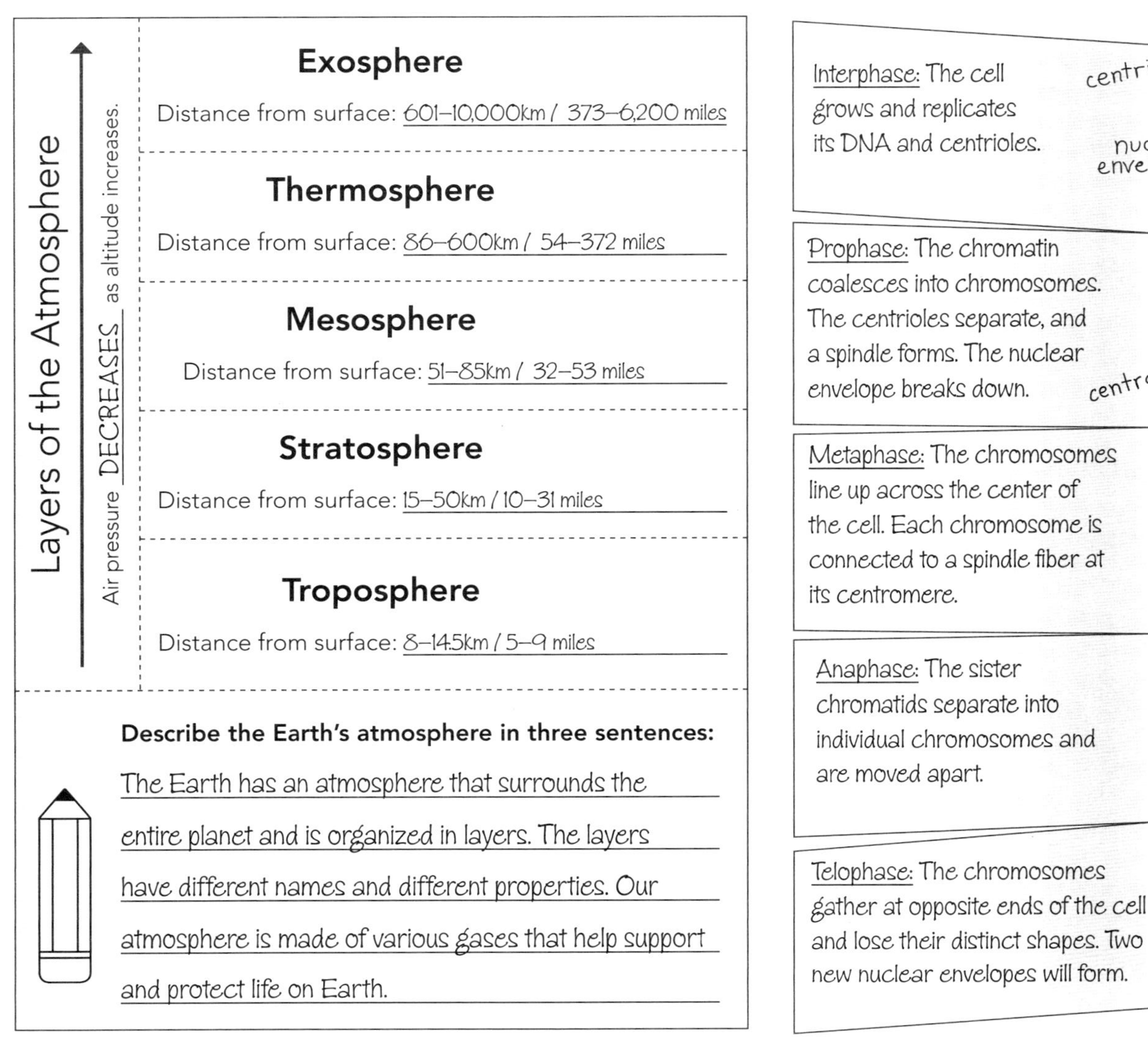

Figure 5.5: Student sample foldable activity.

the actual 3-D cutout of the foldable allows them to focus their energy on the content, not the creation of the tool. Many teachers allow students to use foldables on assessments and during in-class discussions, which gives both English learners and students who qualify for special education a starting point. Because students can manipulate foldables, they are a creative and versatile way to meet the specific needs of *all* students.

For students who have already mastered the science content, ask them to make foldables that draw larger connections to other units of study, or have them apply specific cross-cutting concepts as they are outlined in the NGSS Lead States (2013).

Synthesize Sources and Connect to Prior Knowledge

When students are working with different texts in science class, they have to navigate different modes of information. There are textbooks, lab reports, charts, graphs, and numerous other texts that students need to synthesize to draw larger conclusions. This strategy provides the explicit instruction many students need around how to pull information from various source materials to demonstrate comprehension of science content.

How to Use

When students need to navigate between two or more different pieces of source material, you can provide explicit guidance on how to synthesize the readings. For example, if biology students are reading multiple articles about environmental effects on different species, help students make connections between the science content (from prior knowledge or instruction) and its application (to the articles) by providing a prompt, such as, *Using evidence from the articles, what claim could you make about trait expression due to environmental factors?*

Students then write a well-developed paragraph that takes elements from multiple texts to demonstrate a larger understanding of the scientific concepts. This works with any two or more texts. We encourage you to use this strategy with nontraditional texts—media, data tables, lab manuals, and so on. It's important not to assume your students can do this independently. Synthesis is a skill that has applications well beyond the science classroom. We owe it to our students to model and teach them how to synthesize.

Adaptations

One adaptation to this strategy in the science classroom can be using two texts of the same type when beginning to teach this strategy. For example, take two graphs and have students use both graphs to draw two conclusions or thinking points. Then follow up with two different text types on the same subject (one graph and one passage from a textbook), and have them draw a conclusion about these texts. Gradually move into the varying text types. This helps to scaffold the strategy and teaches the importance of synthesizing data, materials, and texts.

Formative Quick Checks Using Online Quizzes

When the goal is to make sure students are reading and picking out the most relevant information in a text, it is easy to check their progress with a short, simple, formative quick check with online resources like Kahoot! (https://kahoot.com) or Quizizz (https://quizizz.com). Students and teachers can quickly receive feedback about whether the notes and information that they have obtained are important, accurate, and relevant.

How to Use

Read the text in advance and create a multiple-choice assessment based on the essential information you and your team feel is important to comprehend and retain. Access one of the online quizzing platforms, and create the quiz. After students have completed the assigned reading, allow students to use their prereading and during-reading notes on the quiz. If a student's notes clearly recorded the essential information, he or she should perform well on the quiz.

A quick reflection after the quiz will provide immediate feedback for the students regarding their ability to read, comprehend, and identify the essential information. Combining the students' scores by requesting their responses to follow-up questions, such as *Did you have too few or too many notes?* allows them to evaluate their reading performance.

Adaptations

When using the online resources, it helps students to see their answers compared to their classmates'. One adaptation for both students learning English and special education students is to have them work with a high-proficiency partner the first time. This allows the students to build confidence and work collaboratively to build knowledge. In addition, students can also do a quick check-in with a partner over the specifics of their notes. They can review each other's notes and add any information they feel is missing prior to starting the online quizzes.

Postreading Checklist

Another quick, formative check involves asking students to complete a graphic checklist to receive feedback on how well they understood the reading. By completing the checklist, you and your students can quickly and visually identify reading success. As an added bonus, students will have the checklist to use for studying purposes.

How to Use

When appropriate, create a chart based on the content of a reading. The example in figure 5.6 is based on comparing and contrasting two concepts covered in a text. Readings that have compare-and-contrast, cause-and-effect, and other similar structures work best. After students complete the reading, allow them to use their notes to complete the chart. Providing students with an answer key following the exercise allows them to easily gain feedback on whether their notes and reading comprehension were sufficient. Teachers can also monitor how students are self-assessing and provide more instruction if needed. Student reading and notes can be easily assessed by a quick reflection once students are finished and furnished with a key. Students will know whether their reading was sufficient.

Adaptations

This straightforward strategy provides a quick way to check in on students' understanding and is easy to use with and adapt for students learning English or students who qualify for special education. For example, adding page numbers to the characteristics column can provide a more specific reference for students to focus their effort on finding critical text. The chart itself also provides a structure for reading and narrows the referencing for the students.

What Does It Say? What Does It Not Say? How Does It Say It?

Literacy expert and English teacher Kelly Gallagher (2018) created this strategy to develop close-reading skills in the English classroom. We have had success applying his strategy in disciplines beyond the English classroom. The strategy requires students to think about the literal, inferential, and structural elements of a text, and it works very well with small-group or whole-class discussions.

Name Sal Vinero Class Accelerated Biology Date 10/30

Use your reading notes from eText section 9.3 (pp. 262–265) to complete the fermentation comparison table by placing an X in the box related to the characteristic listed in each row. Be sure to answer the note-taking reflection questions to improve your note-taking skills in the future.

9.3 Fermentation (eText pp. 262–265)

Characteristic	Alcoholic Fermentation	Lactic Acid Fermentation
Occurs after glycolysis	X	X
Is an anaerobic process	X	X
Produces ethyl alcohol	X	
Produces NAD+	X	X
Produces lactic acid		X
Produces CO_2	X	
Starts with pyruvate	X	X
Occurs in yeast	X	
Occurs in human muscle cells		X
Used to make bread dough rise	X	
Makes wine and beer	X	
Makes cheese, yogurt, sour cream, pickles, and kimchi		X
Occurs in the cytoplasm of the cell	X	X

Reading notes reflection: (1) Approximately what percentage of your notes were helpful and were used to answer the assessment items? (2) How could you improve note-taking for next time?

1. 90% because there were some characteristics on the checklist that I missed, but most of them were covered.
2. I think I did pretty well, but I might organize my notes into different categories to better understand how different processes work.

Source: © 2016 Amy Inselberger and Dan Argentar. Used with permission.

Figure 5.6: Postreading checklist to compare and contrast concepts from the text.

How to Use

After reading a text, have students identify the answers to three questions.

1. **What does the text say?** This refers to what the text says literally. We often tell our students to identify something they can put their finger on—something that the text literally states. What makes it important?
2. **What does the text not say?** This requires students to make an inference about something that is implied in the text. It could also be important information that the text doesn't include but that the student wants to know more about or feels is relevant to the reading.
3. **How does it say it?** This requires students to explore the structural elements of the text. For example, a student may explore how a table reveals trends from a lab experiment, or how a graph's slope demonstrates a connection to a scientific concept.

After students identify one detail for each question, they can share their work with the whole class or a small group. Because they have had time to think, all students should be equipped to participate in the discussion, and the class can build understanding as students closely analyze the text with a literal, inferential, and structural lens.

Adaptations

One adaptation to this strategy is to provide one of the three sections to the students. So, as the teacher, you give them an example of either what the text says, what the text doesn't say, or how the text says important information. Giving this extra detail helps structure the thinking for the struggling readers, narrows their focus for the other two sections, and provides a starting point for their thinking.

Targeted Entrance and Exit Slips

When determining whether a student has understood a text, one of the more tried-and-true formative assessments to use as a postreading exercise is an entrance or exit slip. These low-stakes check-ins allow you to take inventory on who is mastering the content or skills of a lesson or reading assignment. We encourage adding opportunities for self-assessment that allow students to see clearly the criteria for how you will evaluate them.

How to Use

Provide students with a targeted writing task as they enter the class or before they leave class that requires them to articulate an understanding of a lesson or reading task and turn in their written response. For example, after a reading on cellular respiration, you might ask students to write on a notecard their answer to the prompt, *In your own words, without using notes, explain cellular respiration as best you can.* One way to ensure that students have truly mastered the material is to ask them to explain a concept in their own words. This helps avoid the temptation to rewrite or copy directly from the text. You can also encourage students to extend their learning by applying the content to a broader context. The possibilities for questioning are endless, and it is a simple yet effective method to take the pulse of the classroom. For entrance slips, students may have opportunities to compare their work with peers or exemplars to help them prepare for the lesson that follows. For exit slips, note common patterns in students' responses prior to the next class meeting, and adjust instruction accordingly.

Adaptations

For both students learning English and students who require special education, you can tailor your questions to a specific student's needs to assess his or her level of comprehension. You can use these slips for either literal or inferential questioning. For questions that ask for a writing response, you can provide the students with a couple of vocabulary words you want them to use in their response. This provides a structure for their thinking and gives them some prompts to start their writing.

Considerations When Students Struggle

Students need to practice postreading strategies and thinking in order to develop their analysis and critical-thinking skills, and there are a number of factors that can lead to struggles for students in this area. For example, struggles may occur for students with the format of a strategy or with the level of thinking it requires. Here are some additional questions to consider when you encounter students who are having a hard time applying postreading strategies.

- How might you ensure students have the literal information required to think more deeply?
- How might you push students at different levels to deepen their analysis?

- How might you meet groups of students at different thinking levels?
- How might you tailor the strategies in this chapter so that students can use their strengths for similar tasks or think in a way that is developmentally appropriate for middle school, high school, or college?

As teams begin to explore these questions, there are a number of factors to consider. It's important to remember that postreading instruction and strategies are part of the reading process. Teams should consider the effectiveness of prereading and during-reading instruction. If students struggle in the postreading stage, it may be an opportunity for teams to reflect, revise, or reconsider the instruction for prereading or during reading.

Remember that the strategies we provide in this book are designed to help students read, write, and think like scientists. Teams should collaborate around ways to adapt the strategies to meet the specific curriculum's requirements as well as the needs of students. Consider adapting vocabulary and terminology so that they align with what students see in the texts and hear in the classroom.

Review the strategies in this chapter. How might you help students who struggle to make progress?

Considerations When Students Are Proficient

There are a number of team considerations for students who have demonstrated proficiency throughout the postreading process. Teams working within a PLC should collaborate to make sure that all students are challenged appropriately. Here are some considerations for differentiating instruction for proficient readers.

- Students can step into a mentoring or coaching role, helping fellow students to craft questions or clarify understanding. Teaching the material to others empowers students, while solidifying and extending their own knowledge.
- Teachers can pair students with other proficient students to extend their learning through supplemental texts, nontraditional texts, or research. Self-directed learning challenges students to develop their own inquiry and extensions based on their personal interests and questions.

- Proficient students can generate their own assessments or questions (review them for appropriateness and accuracy!) to challenge less-proficient classmates as their peers work to practice and develop their own proficiency.

Wrapping Up

Moving beyond the literal reading and understanding of a text takes work. Students do not always default to deep postreading analysis. Encourage them to be more thoughtful about how the material and the application of their scientific knowledge will benefit them as they move through each level of schooling.

Collaborative Considerations *for* Teams

- What postreading strategies can teams adopt for use in the classroom? Consider the following.
    - How will the team implement these strategies?
    - What adaptations may be necessary for the team, task, or students?
- How can postreading strategies provide formative data for teams to impact future instruction?

CHAPTER 6

Writing

One day during a spring semester, we caught wind of some discussions that were happening in the teachers' lounge between one of our English teachers and one of our science teachers. They were talking about their respective approaches to teaching their students how to write summary paragraphs. The English teacher had students organizing their summaries with a main idea followed by supporting details and explanations. The science teacher explained how the students were writing science content organized into three parts: (1) context, (2) summary, and (3) explanation. That moment spurred a larger conversation among content-area teachers interested in literacy about the commonalities we expect of our students in each discipline when writing. It was from that conversation that our first real progress toward a common writing literacy language was born.

This chapter highlights the ways that you and your team can incorporate effective writing instruction into the science classroom. First, we tell how teachers we have worked with were able to enhance their writing instruction by collaborating with teachers in other disciplines to create a common schoolwide writing vocabulary aligned with language in the CCSS and the NGSS for grades 6–12. We then highlight the NGSS that align with essential literacy skills and offer concrete writing strategies that you and your team can implement to enhance writing instruction in all science classrooms. Finally, we offer considerations for addressing students who continue to struggle and those showing high proficiency.

Collaboration Around Consistent Language

Before the start of the following school year, as literacy coaches, we seized the opportunity to open up a conversation about writing. We sought interested volunteers from all content areas to share their thoughts about writing. The response was robust, as we found teachers from almost all content areas willing to share. Teachers were committed to helping support students' skills in this area.

The first meeting consisted of teachers from different content areas sharing their writing expectations, structures, and samples. Subsequent meetings involved teachers from different backgrounds examining the different types of writing and finding commonalities. From those conversations, it became clear that, as a school, we should be consistently using the same terms, such as *claim* (instead of *thesis*, or *hypothesis*), *subclaim* (instead of *topic sentence*), *evidence* (instead of *quotes*, *examples*, or *data*), *reasoning* (instead of *justification*, or *elaboration*), and *conclusion*. In this way, as students move from class to class and discipline to discipline, they are able to hear the same language from all of their teachers. Yes, there are unique qualities to writing in a science class, and students need to learn those differences, but students are able to make stronger connections with what "good writing" looks like in different disciplines when they see how a term (such as *evidence*) is used in different academic contexts. Having teachers discuss this and model it in each of the disciplines clarifies what it specifically looks like in each class. Further, using the same terminology and words when instructing helps students make stronger connections and understand the expectations of what they are writing.

Toward the end of the fall semester, we met again as a group to reflect and discuss results with anyone who had tried to use the common writing language we developed previously. This conversation was overwhelmingly encouraging. Mathematics, social studies, and science teachers reported increased success with student writing once they started using the common language. For example, when science students understood that restating their hypothesis could work as their *claim*, they had a clearer idea about beginning their science conclusion paragraphs. When they realized that their data would serve as specific *evidence*, while their justification was their *reasoning*, their writing became more succinct and clear with a purpose.

Admittedly, content-area teachers did not just come up with the terms *claim*, *evidence*, and *reasoning* on their own. In fact, those terms are key elements of argumentation, and they are embedded in the CCSS for English language arts (NGA & CCSSO, 2010). One of the first times educators in our school encountered the common argumentation language was in meeting with educator and author Katie McKnight in September 2014. She brought to our attention the work of former teacher Eileen Murphy and her colleagues at ThinkCERCA (n.d.a), a literacy courseware company (CERCA stands for *claim*, *evidence*, *reasoning*, *counterargument*, and *audience*). After that 2014 meeting, different groups of teachers and curricular teams began using these terms, and after a few years of the ideas organically filtering through different divisions and collaborative teams, our content-area teachers decided to loosely adopt the terminology.

Some teams made the decision to change some of the language they had been using, such as changing *thesis* to *claim*, and those that committed to using the language saw immediate benefits. One, they realized that the language aligned with the language used in standards such as the NGSS or the CCSS. Two, they shared anecdotal evidence that their students were writing more effectively because of the common language they heard throughout the school day. What started as a loose adoption transformed over time into a schoolwide writing vocabulary. The writing vocabulary still has not been made official or mandated from leadership, but there are only a small handful of teams in the building that use unique language; it's a work in progress the PLC continues to hone.

According to ThinkCERCA (n.d.b) research, "teaching students how to make Claims, support their claims with Evidence, explain their Reasoning, address Counterarguments, and use Audience-appropriate language is the most effective way to improve achievement on assessments and prepare students for postsecondary life." With research and logic behind us, we were happy to facilitate the proliferation of this common language.

This is all to highlight that, for many science teams working in a PLC, explicit instruction around writing may be new. Other teams may have been working with writing instruction for some time but lack a common instructional approach. Some teachers may emphasize different elements of writing instruction over others. Therefore, an important first step for a team of science teachers in a PLC is to reach consensus around the first critical question: What do we want students to know? Science teams will need to come to consensus around what every student should be able to do as a writer within the core curriculum. Subsequently, teams can plan units with writing instruction in mind. It's important to start small. Consider how your team can implement one writing experience per unit. From there, teams can expand or adjust their writing experiences as needed.

How can you implement the ThinkCERCA model or other argumentation tools in your classroom?

NGSS Connections

With a focus on writing and argumentation, the grades 6–12 NGSS Science and Engineering Practices clearly expect students to create arguments based on evidence. Whether the arguments are oral or written, the standards require higher-order critical-thinking skills as students prepare claims, evidence, and reasoning. To see some of this language in the standards, we boldfaced relevant words in figure 6.1. You might notice that there are more nouns highlighted in the following standard language than in previous chapters focused on reading strategies. This is mostly to reflect the writing skills that students will be expected to demonstrate throughout their education in the science classroom.

Construct and present oral and **written arguments** supported by empirical **evidence** and scientific **reasoning** to support or refute an explanation or a model for a phenomenon or a solution to a problem. (MS-PS2-4; MS-PS3-5; MS-ESS3-4; MS-LS1-3; MS-LS1-4)

Integrate qualitative scientific and technical information in **written text** with that contained in media and visual displays to clarify **claims** and findings. (MS-PS4-3)

Communicate scientific and technical information (e.g., about the process of development and the design and performance of a proposed process or system) in multiple formats (including orally, **graphically**, **textually**, and mathematically). (HS-PS2-6; HS-PS4-5; HS-LS4-1; HS-ESS1-3)

Construct an oral and **written argument** or **counter-arguments** based on **data** and **evidence**. (HS-ESS2-7)

Science **arguments** are strengthened by multiple lines of **evidence** supporting a single **explanation**. (HS-ESS2-4; HS-ESS3-5)

Source for standards: NGSS Lead States, 2013.

Figure 6.1: Writing and argumentation connections in the NGSS.

Strategies for Supporting Students in Writing

There are so many different ways to get students to produce thoughtful writing. Along with utilizing a common writing vocabulary, using mentor texts (published writings that students can emulate), creating models based on students' efforts, and cowriting can all lead to wonderful results over time. More often than not, strategies come down to modeling and practice. When possible, work with other teachers to share samples, practice ideas, and give feedback. When you collaborate with different teams within a school community to share the ways you all teach and assess writing, you can make powerful connections around the ways that all of your students engage in different writing tasks in different disciplines. The following strategies and differentiation options for students learning English and students who qualify for special education, created in conjunction with our science teachers, will help you and your team build students' scientific writing competency.

Two-Quote Paragraph Template

Sometimes, structured, formulaic writing is the best way to teach students how to start writing (and thinking) like scientists. This tried-and-true method is an effective tool to lay down strong writing foundations at the beginning of a course. Our science teachers have found that using what they call the *two-quote paragraph* helps students make connections to writing instruction in other disciplines while building important science argumentation skills required in lab reports and the NGSS. Called a two-quote paragraph because it asks students to include two quotes to support their argument, this strategy offers students a template for constructing an argumentative writing paragraph. The goal is to eventually break away from the template as students internalize the writing structure; nonetheless, providing students with a sturdy structure for writing helps them ensure they are using strong logic and reasoning in an organized manner.

How to Use

This strategy is most effective when teaching the basics of scientific argumentation. We recommend using this early in the school year when students are learning how to construct lab conclusions or do other argumentation tasks. Students will use a template, like the example in figure 6.2 (page 118), to ensure they are incorporating all required argumentation elements: claim, evidence, and reasoning. (See page 158 for a blank reproducible version of this tool.)

In this example, tenth-grade chemistry students are writing a paragraph to practice writing scientific conclusions following an experiment around humans' production of carbon dioxide. It provides a clear, sentence-by-sentence structure for students to follow, and you can easily adapt it to meet the needs of the students and the writing task. Is it formulaic? Yes, but that's the point. It helps students learn what good writing looks like. Students often ask, "How long does this have to be?" While the template is flexible, it does help to provide guidance on where students need to elaborate on their ideas.

Sentence Number	Type of Sentence
1	**Topic sentence:** State your *claim* (purpose of the text). If a person's breathing rate increases, then the amount of carbon dioxide exhaled decreases.
2–3	**Lead-in:** Establish a context for the upcoming evidence, or explain a little about what you are claiming. You might define vocabulary or add context to further your idea. In the procedure to conduct this experiment, scientists found ten male volunteers of approximately the same height and weight. The male volunteers were hooked up to a breathalyzer that was able to measure the amount of carbon dioxide exhaled by the male volunteers. The scientist instructed the volunteers to breathe only when told to by a flashing light. The light flashed three times during the first three minutes, eleven times the next three minutes, and so on through 14 breaths a minute. The exhaled level of carbon dioxide during each three-minute trial was averaged and added to a data table.
4	**Quotation or evidence:** Make sure you choose one of the best quotes or pieces of evidence from the text that will support your claim. As shown in the data, persons whose breathing rate was 10 breaths per minute produced 42 mmHg carbon dioxide.
5–6	**Justification or reasoning:** Now explain why your evidence is important. How and why does this evidence help support your claim? Explain the reasoning behind why your evidence proves your claim. This was the original data point taken early in the experiment. This provided a baseline that we were able to compare when the breathing rate increased.
7	**Quotation or evidence:** Make sure your second quote or piece of evidence supports your claim as well. Make sure you transition carefully into the evidence; don't just drop it in the middle of the paragraph with no sentence setting it up. A person who has 14 breaths per minute produces 28 mmHg carbon dioxide.
8–9	**Justification or reasoning:** Now explain why your evidence is important. Why does it help to support your claim? Explain the reasoning why your evidence proves your claim. There is a clear decrease in the amount of carbon dioxide between these two data points. There was a drop of 14 mmHg carbon dioxide produced when the breathing rate increased 4 breaths per minute.
10	**Return to the claim (concluding sentence):** Return to the claim. This last sentence should relink your evidence and ideas back to your primary claim. The significance of this experiment is to see if a person produces more carbon dioxide when being active or staying still.

Figure 6.2: Sample two-paragraph template.

Adaptations

When thinking about how to adapt this model for students learning English and students who qualify for special education, the first thing to consider, as mentioned previously, is to model the process multiple times. Modeling the strategy gives students who may struggle to immediately process spoken language a clearer opportunity to know what the expectation is and what is required of them. Having the students work with partners also helps clarify ideas that may not be as easy to comprehend. In addition, think about providing specific words you expect the students to use in their writing in each of the template boxes. You can also give the students the claim statement and then have them work through the remainder of the strategy. Ultimately, the goal is for the students to complete the paragraph independently, but this may take a few extra steps to help students learn and understand the process.

Highly proficient students who have mastered paragraph organization may no longer require the use of the template. This is the goal for all students, so challenge them to produce the same quality and level of work without a template. For those who take longer to demonstrate mastery, have them continue to use the template as needed.

Evaluate and Support Claims With Evidence

This strategy works best for laying the foundations for argumentative writing skills in the science classroom. It asks students to evaluate claims and then independently identify evidence that supports the most effective claim statement. This is a powerful formative assessment in that students are able to see a range of claim statements, and then they are required to find evidence that best supports each claim. By scaffolding instruction to help students first identify what an effective claim looks like, followed by identifying the most appropriate evidence, you are able to break down the writing and thinking process into separate stages.

How to Use

Provide students with three or more claim statements that range in effectiveness, and ask them to evaluate the claim statements and select the best one. (See figure 6.3, page 120, for an example; see page 160 for a blank reproducible version of this tool.) After students make their selection, do a formative check and discuss why students made specific selections, or allow students to begin to select evidence for whichever claim they selected. If students chose the most effective claim, they should have little difficulty finding appropriate evidence. If students chose a less

Why Are Some Bond Angles Different When There Are the Same Number of Domains?

Three students develop claims to explain the difference in angles for the three molecules shown in the following figure. Read each of their claims. Put a checkmark next to the student you feel has the *best* answer. After choosing a student, read pages 126–127 in your textbook, and find *at least two* pieces of evidence to support your chosen student's claim.

Claim 1: Nicola Amerigo says the number of hydrogen atoms attached to the central atom determines the size of the bond angle.

Claim 2: Graziano Pelle says the bond angle is determined by the number of unpaired electrons attached to a central atom. ✓

Claim 3: Andrea Pirlo says the farther to the left on the periodic table the central atom is located, the larger the bond angle.

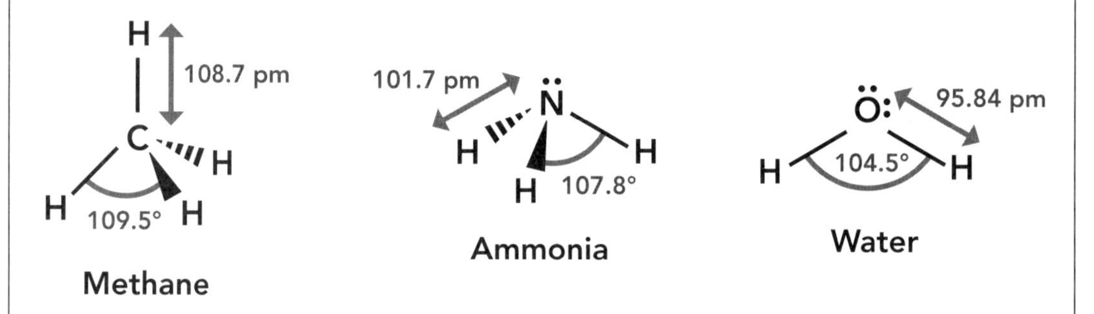

Evidence	Page Number	How the Evidence Supports Your Chosen Claim
"The number of lone electrons around a central atom decreases the bond angle."	126	Since there are the most lone electrons around the oxygen atom in water, it has the smallest bond angle.
"Water has lone electrons, and methane does not; therefore, their bond angles are different."	127	Bond angles differ when comparing two molecules when one has lone electrons and the other does not. This supports Pelle's claim.

Figure 6.3: Evaluating claim statements and supporting them with evidence.

effective claim, they might struggle to find evidence, which may require them to rethink their original selection. Require students to identify multiple pieces of evidence and provide reasoning for each selection.

Adaptations

You can adapt this strategy for students learning English and students who qualify for special education by allowing them to begin to locate evidence for whichever claim they selected. If students chose the most effective claim, they should have little difficulty finding appropriate evidence. If students chose a less effective claim, they might struggle to find evidence, which may require them to rethink their original selection. For students who continue to struggle, it's okay to provide them with just the correct claim to help them practice evidence selection. As the students progress, you may add in a second claim and then another to build the challenge. You may also reverse this strategy for students. You can provide the evidence for them and have them then come up with the claim. This helps them process the information and try to fit it all together without having to come up with the claim evidence. Eventually, students will be able to use this strategy without these adaptations, but scaffolding it, at first, will help them work with the ideas.

For students showing high proficiency with this strategy, consider providing a claim that may seem viable, but is provably false. Have students work to *disprove* the questionable claim in addition to proving the viable claims.

Color-Coded Paragraphs

As we stated earlier in this chapter, when students are learning how to write argumentative paragraphs in science, they frequently ask, "How much do I need to write?" Additionally, they need guidance to ensure they have all the necessary components of an effective argumentative paragraph. Color-coding drafts of student writing is an effective visual strategy to help students see how each component of the argument contributes to the larger whole. Students will have a clear visual indicator for every component in the paragraph. By comparing their colored paragraphs with exemplar or peer writing, students will be able to assess their own writing.

Expectations for the product students produce will vary based on the writing task, so teachers should clearly indicate to students what an appropriate ratio looks like based on the assignment. As for a balanced end product, the students should have fewer sentences for the claim section, and more written for the evidence and reasoning. Again, this can vary depending on the individual assignment.

How to Use

Assign a color for every required component of a writing task (for example, *claims* are red, *evidence* is yellow, *reasoning* is blue, and so on). Students then use a highlighter or a word-processing application to highlight the entire paragraph, applying the designated colors to each paragraph component. Students can compare their own highlights with an exemplar or peer's work to ensure that the quantity and ratio of sentences for each of the areas (claims, evidence, reasoning, and so on) of their writing are sufficient. Students can work together to discuss differences of opinion in how they assessed text. This is not an effective strategy for assessing the *quality* of the writing, but it does ensure that students have the necessary components. It's also a good strategy to use as students begin to break away from the two-quote paragraph template (see page 117).

As an example, figure 6.4 shows an excerpt from a student's lab conclusion for testing how a person's breathing rate affects the amount of CO_2 he or she exhales. The sample is annotated as follows: the claim is shaded in pink (light gray on this page), the evidence is shaded yellow (medium gray), and the reasoning is shaded in blue (dark gray). In this instance, the student is able to see that his writing has all of the necessary elements for this writing task and that it has an adequate amount of writing for the given task. (Remember, the amount of writing necessary will vary with the task.)

If a person's breathing rate increases, then the amount of carbon dioxide exhaled decreases. To conduct this experiment, scientists found ten male volunteers of approximately the same height and weight who were hooked up to a breathalyzer that was able to measure the amount of carbon dioxide exhaled. The volunteers breathed only when they saw a light flash. During the experiment, a person whose breathing rate was 10 breaths per minute produced 42 mmHg carbon dioxide, whereas a person who had 14 breaths per minute produced 28 mmHg carbon dioxide. As the data shows, my hypothesis was supported because I said as the breathing rate goes up, the amount of carbon dioxide exhaled will go down. Since the rate was controlled by the light, and all of the subjects experienced a decrease in exhaled CO_2, we assume the metabolic process is consistent.

Figure 6.4: Color-coded excerpt from a student lab conclusion.

Adaptations

This visual-based strategy is especially helpful when used with students learning English and students in special education because it gives them a clear and distinct key to understand what they are writing and make sure they are hitting all of the components of their writing in their science reports. If students are overwhelmed by focusing on multiple elements of their writing, teachers can scaffold the instruction by having them highlight one element at a time (focusing on one skill at a time).

Similar to with previous writing supports, like the two-quote paragraph, students who have consistently demonstrated proficiency with writing organization should not have to use this strategy. As more students reach mastery with writing organization, they can begin to break away from some of these scaffolded supports.

Considerations When Students Struggle

Writing is difficult. There are so many issues with writing that students will undoubtedly need help and support mastering it in the science classroom, from getting started to grammar to content to organization. We know how incredibly frustrating it is for you (as well as students and parents) when students know answers but simply struggle to articulate ideas in writing. With that in mind, here are some additional questions to consider when you encounter students who are having a hard time writing.

- How might you break down tasks to allow students to focus on one idea at a time?
- How might you help students write first and worry about editing later?
- How might you find time to confer with students one-on-one to help them before they write? Would it benefit a student to work one-on-one with you as he or she writes in the moment? Or for you to work with him or her one-on-one after writing? This may vary depending on the assignment. Working with the student during either of these periods provides a model for future work and ultimately for the student to complete it independently.
- How might you pair students together to build on each other's writing strengths?

Because writing is a complex task, teams will benefit from identifying one or two concrete skills to focus on for collaboration and instruction. As teams (and students) begin to master and build on those skills, they can refocus attention on other skills, coming to consensus on instructional strategies like the ones outlined in this chapter. As teams experiment with different instructional approaches to complex writing tasks, each team will continuously increase its capacity to support students as science writers.

Use the work of proficient students, ones who do not need templates or visual indicators to find success, as examples for other students to view. The instructor can also use this time for the students to work with each other to review their write-ups and make sure all components of claim, evidence, and reasoning are in the write-up.

Review the strategies in this chapter. How might you help students who struggle to make progress?

Considerations When Students Are Proficient

Students who have demonstrated proficiency in writing still have room for growth. Here are some suggestions to support students who are ready to take on more complex writing tasks.

- Encourage proficient writers to increase synthesis skills by asking them to embed multiple sources into their writing. This allows students to make connections within and beyond science units.
- Experiment with different sentence variety, perhaps encouraging students to emulate published science writers.
- Increase the use of more technical science vocabulary, as relevant to the unit of study.

Wrapping Up

Reading and writing go hand in hand. More often than not, growth in one area can lead to growth in the other. Therefore, it is important to keep students practicing reading and writing in science classes. Conference, model, and share student samples with them to show students that writing is a part of science and to help each student learn what he or she can do to grow as a writer.

Collaborative Considerations *for* Teams

- Is your science team equipped with a writing vocabulary that aligns with the language from your school's standards, such as the CCSS or NGSS?
- Are there opportunities for your science team to collaborate with teams from other disciplines? Who can facilitate those conversations?
- Review the writing strategies in this chapter. Consider the following questions.
  - What strategies would fit your team's writing instruction needs?
  - Can the team develop mentor texts and models to provide exemplars of effective writing?

CHAPTER 7

Assessment

When the time came for Cami's team (see page xvi) to address assessment in our collaboration, their frustration was clear as members started to feel overwhelmed: "Wait! Now I have to assess my students like an English teacher? I can't possibly make time to assess both my content and *your* reading skills." However, the solution to these concerns is simple. In fact, the assessment of reading and the gathering of knowledge from texts are intertwined. All we had to do was re-examine the NGSS to see the literacy skills embedded in the Science and Engineering Practices. The standards were *already* asking the teachers to assess content through literacy practices. For example, when the NGSS ask for students to be able to *analyze data* or *support a claim*, teachers are tasked with assessing both the content usage and how the students are analyzing or supporting their knowledge.

We then had a lengthy discussion with the team on what exactly we were trying to assess. Were we assessing content knowledge? Literacy skills? Ultimately, the answer was *both*. If we wanted students to grow and succeed, we needed to give them feedback, which meant assessing their work and holding them and ourselves accountable. Together, we set out to find ways to make most literacy assessments quick and seamless so the science teachers could focus on content. Logically, the science teachers wanted the content to drive the student evaluations.

We found, together, that through quick formative checks and feedback, we could assess student literacy work and produce growth. Sometimes the assessments came in the form of warm-up activities like the Kahoot! quizzes mentioned in chapter 5 (page 105). Sometimes the assessments were simple visual scans of whether students ultimately agreed or disagreed with a statement on the anticipation guide activity mentioned in chapter 3 (page 58). And sometimes, teachers had to embed the assessments in more developed thinking activities like lab write-ups that took more time but, in the end, allowed students to more clearly articulate content targets.

The challenge (and the fun) of working as a collaborative team ensured that, from start to finish, activities were useful for both literacy *and* assessment purposes—and teachers felt comfortable implementing them in their classes not only because they created them together but also because they knew the team would also work together to assess the resulting data and use them to further instruction.

There are seven considerations to keep in mind when approaching assessment, including (1) understanding the role of literacy-based assessment in the science classroom, (2) matching text complexity and reader capacity, (3) monitoring student perceptions, (4) collaborating to create assessments, (5) using rubrics as assessment tools, (6) providing timely and effective feedback, and (7) analyzing and applying data. In this chapter, we break with the format we established in previous chapters to focus specifically on each of these considerations.

Understanding the Role of Literacy-Based Assessment in the Science Classroom

Assessment is a natural component to the classroom and always has been the primary source of identifying a student's understanding of content. Assessments drive many elements of learning within our schools and beyond, including determining final grades and influencing college admissions, which often leads schools, media, and society to focus heavily on this educational tool. The importance that assessments bear means that your school leadership and community stakeholders expect you, as an educator, to be up to date on the most recent assessment trends and subscribe to some sort of methodology. This task can be daunting; whether it be assessment *for* learning or standards-based grading, there is always a new idea out there.

While there are many assessment philosophies, the common element among most of them is using assessment data to inform instruction. If we are not using assessments to better identify our next steps as teachers, we are misusing the data they produce. After reviewing students' assessments, you may respond by giving a follow-up lesson or by reteaching a concept the next day, or a more long-term response involving curriculum revision may be appropriate. Next steps will depend on the purpose and format of your test. Typically, most assessments fit into one of three categories.

1. **Formative:** These are in-class assessments that you specifically craft to inform yourself and your students of their progress toward mastery. These are not final tests, and there should always be another assessment

opportunity. These tests may carry score weight, but they are not the last opportunity for students to demonstrate their knowledge of a target, standard, or task. Note that depending on your school's testing philosophy, formal progress monitoring (between benchmark periods) may also fall within this category.

2. **Summative:** These assessments are less frequent and serve as an end-cap to teaching and practicing a concept. You may administer summative assessments in class or even as part of a larger context, like course final exams. That said, like with all team assessments, we encourage you to use results from any summative assessment to drive continuous improvement in teaching and learning.
3. **Benchmark:** Benchmark assessments are given at specific intervals throughout the school year and aim to capture student academic growth from a holistic perspective. Many teachers use benchmark assessments to gauge students' knowledge at the start of a semester, school year, or high school career, and then again at the end. Some include a benchmark assessment at the midpoint of these as well. These data are typically for tracking students' learning progress and not for determining a student's grade in a particular course. These assessments are often group administered and nationally normed, meaning that the tool delivers a percentile comparing each student to others in the same grade or of the same age within a statewide or national pool of students representing all student populations.

As your team approaches the work of literacy assessment within the science context, keep the four critical questions of your PLC in mind. These will guide your team discussions and point to your next steps. You may apply these questions holistically to an assessment or to each assessment component. As you do so, trends will emerge, once again pointing you toward your next step, which may range from reteaching a specific lesson in the classroom to reteaching with a small group of students to engaging one-on-one with a specific struggling learner. Additionally, your findings may point you toward your next level of inquiry in the form of a deeper-dive assessment.

Matching Text Complexity and Reader Capacity

When crafting literacy-based assessments for the classroom, it is vital that you design with text complexity and reader capacity in mind. First, you must know the readers in your classroom and their skill sets. Administering an assessment that provides a Lexile measure for each student is a practical way to get this information. Typically, these assessments are intended to be administered to a large group of students at one time and are often computer-based, allowing scores to be available immediately. Given RTI and MTSS have become common practice across the United States, many schools are already conducting this type of benchmark assessment (Buffum, Mattos, & Weber, 2012; National Center on Intensive Intervention, n.d.); however, time and again, we have noted that the results of such benchmarking often go unused.

To ensure schools use these results effectively, leadership teams can consider the following questions.

- Does the school conduct benchmark literacy assessments?
- When does benchmark testing occur?
- Which student populations participate?
- Where are the data housed?
- What happens with the data?
- How can these data be shared with core departments?

Note that anyone can start this conversation. Bringing these questions to your leadership is simply one way to initiate and request the support that your team needs and also prompt your leadership to partake in reflection.

Because all content-area teachers are teachers of reading, it is imperative that you are able to access students' literacy data. These data will help you scaffold instruction and select *materials* and *texts*. Additionally, they will help teachers gain a better understanding of why students are striving or struggling. The goal is to understand each student's independent, instructional, and frustration reading comprehension levels, as follows.

- **Independent:** This is the level of text that a student can navigate independently with fluency and solid comprehension. The student does not need any supports or scaffolds when engaging with this text. A student still learns and gains new insights with these texts, but more likely through learning new background knowledge and making inferences, not necessarily by expanding his or her working vocabulary. Matching a student with an independent-level text is a great option for independent reading tasks.

- **Instructional:** This level of text is an appropriate level for reading materials that will be used within the classroom to challenge students. This level fosters literacy-skill growth, if the teacher provides support during the comprehension task. A student's instructional reading range falls within the independent and frustration levels and can often span multiple grades of text complexity; for example, a student's instructional reading range may span grades 8–10.
- **Frustration:** A text at a student's level of frustration is one where the student will have large gaps in understanding due to his or her not having sufficient background knowledge and vocabulary skills to decipher meaning from the text. Employing such a text in the classroom is best reserved for one-on-one or small-group teacher-guided reading with heavy vocabulary and background knowledge frontloading (see strategies in chapter 3, pages 49–54).

The purpose of knowing this information is to use it as you select texts, build your curriculum, and craft your assessments. Clearly, there should be a match in text complexity between lessons and assessments. If you know that you have students who have a sixth-grade instructional reading level, then using a textbook with a tenth-grade reading level would be at their frustration level. Unknowingly, this happens a lot.

As teachers, we have honed and developed our own strong literacy skills, and as we are teaching high school courses, we often assume that our students are on the same track; but the reality is that many of our students cannot work with grade-level materials yet. This reality is sometimes masked by confusion over the content when the root of the issue is weakened literacy skills crippled by poor vocabulary and limited background knowledge. By sticking only with grade-level materials, you are missing opportunities to build the vocabulary and background knowledge that your students will need in order to move *toward* grade-level reading comprehension.

Although many of your textbooks might be well above the instructional reading levels of your students, the solution cannot be to abandon a mandated textbook and summarize it for your students. We have seen this trend in the past as teachers moved away from books and toward slide-based presentations (such as PowerPoint) as a means of text summary. Instead, we encourage you to supplement with interesting, relevant texts that will help build background knowledge and student vocabulary and to frontload challenging text with prereading strategies as scaffolds.

In doing so, you better prepare your students to tackle that challenging textbook in class, with you to support them—not at home, left with their own frustrations.

Monitoring Student Perceptions

Assessing students' perceptions of their literacy skills and practices provides insight into how our students approach text and navigate comprehension pitfalls. In 2018, Katherine's school's literacy team decided that, as part of its school improvement work, the team wanted to survey students to see how they engage in text for academic and nonacademic purposes. We administered a survey in student-friendly language that asked about specific strategies and the scenarios in which they would apply these skills. On a schoolwide level, the data were certainly interesting, but on a department level, the results had more meaning. For example, annotation is something that the English and reading departments viewed as essential to mastering their content, although few students reported that they do this as a during-reading activity. Additionally, most students reported that, when they feel confused while reading a text, they reread. This sounds great, but it made us wonder—Do students know how to reread? Or, are they simply taking the same faulty approach that led to their confusion in the first place? This example led us to the important next steps: we need to model fix-up strategies for our students and scaffold self-remediation skills.

Collaborating to Create Assessments

The work of designing assessments is best done as part of a collaborative team; whether as part of a PLC or not, it is best to work with others on the task of creating literacy-based assessments. Doing so, in some capacity, helps you write assessments that accurately measure student growth with content (learning goals) and process mastery. It is important to assess student growth in both of these domains, as it is really the latter that will help students achieve deeper understandings and move toward proficiency with disciplinary literacy.

The texts you select to use for your assessments should speak directly to your curriculum standards and echo your formative text tasks. In crafting your own reading skill inventories, you will need to make sure that the text-topic needs mirror or extend your course content. Additionally, the questions and analysis tasks should directly speak to the literacy outcomes that your discipline requires.

Make sure that the texts are also within an appropriate instructional range for the students you are testing.

Typically, you will want to design two or three assessments that target the same skills and give them as either a preassessment and postassessment or a preassessment, mid-unit assessment, and postassessment. It is important to give the preassessment prior to introducing any skills to establish baseline data and help determine the level of support students need. You should administer the next assessment after there has been time for thorough instruction, guided practice paired with checks for understanding, and independent application. These assessment results will provide ample discussion opportunities for your team to shape how you teach to your power standards (see chapter 1, page 22).

Know that an assessment does not need to take on one specific format. In fact, a variety of assessment methods will convey to what extent students are able to transfer and transport content-area literacy skills from one context to the next. It is also essential that your assessments are balanced, meaning they should not be all formative nor all summative. Further, although formative assessments should be your primary tool, this does not necessarily mean that all formative assessments should receive a grade; rather, what matters most is that you are continually providing students with opportunities to demonstrate their learning in incremental phases, leading toward summative assessment tasks.

Literacy-based assessments prompt students to employ the same literacy skills they have been practicing, but with a new text that is of a familiar format. For example, if you have been teaching students to identify the main idea of a primary source in a scaffolded fashion, then you should assess this same skill in a similar fashion. If you are teaching struggling readers, you may even want to prompt students to employ the same scaffolds you taught and modeled in class; over time, you can and should eliminate these scaffolds as students gain independence and autonomy as content-area readers. In such a scenario, the strategies we outline in this book can serve not only as your strategy scaffold for teaching a specific literacy skill but also as your assessment of the student's ability to apply that learned skill to new contexts.

Using Rubrics as Assessment Tools

Rubrics have long been a useful tool for teachers to provide students with specific, targeted feedback on their work. When given to students for the purposes of self-reflection, rubrics can serve as a powerful self-assessment tool. There are many different rubric variations based on the specific target or task the rubric wants students to focus on. We've seen it be common practice for rubrics to detail a *minimum* of twenty descriptor boxes, each indicating a level of correctness. Although we sometimes still find these useful for a final summative assessment, our team has moved toward using single-target rubrics as part of the formative and reflective learning process.

The sample rubric in figure 7.1 focuses on a single target. (See page 162 for a blank reproducible version of this tool.) In this case, the rubric details one target on a proficiency scale (level 3 indicating proficiency with the learning target), with specific success criteria, and provides a space for student reflection and teacher feedback. By providing this rubric to students when presenting the assignment or prior to the assessment, you allow students to know exactly the skills they must engage with to complete a task successfully. Additionally, you can have students complete the reflection column prior to submitting their work to add an extra layer of self-editing; alternatively, having students complete it after scoring and teacher comments prompts reflection. Following this with another opportunity to demonstrate skill development on a new practice task or assessment gives students a specific and concrete aspect of the task to focus their efforts on.

Name: Dylan Brooks

Unit: 3 **Targets:** Understand plant cell structure.

Subject: Accelerated Bio—Literacy rubric for reading on plant cells

4	3	(2)	1
I always read and understand a variety of texts by effectively taking notes, asking questions, and applying content. I can apply content to connect knowledge between units and across science disciplines. I am able to help my peers by teaching and evaluating their reading strategies.	I can consistently read and understand a variety of texts by effectively taking notes, asking questions, and applying the content. I recognize when I do not understand and ask specific questions to seek support.	I read and understand a variety of texts by taking notes, asking questions, or applying the content. Occasionally, I recognize when I do not understand, and I may ask questions or seek support. I make some connections between concepts within the reading.	I read for completion of a reading assignment. I cannot identify relationships or concepts within the reading. I do not observe all aspects of the text (skip pictures, graphs, and so on). I do not realize when my understanding stops, and I do not ask questions to seek support.

Success Criteria Teacher circles one: M = Mastery NI = Needs Improvement	**How Well Am I Doing?** Student identifies one strength (+) and one area for improvement (–).	**Teacher Feedback** Teacher circles or highlights successfully met criteria.
Note-taking skills (M) NI	+ I took a lot of notes on cell structure. – I should ask more questions and look up more words.	(Effectively uses notes to break down reading and retain information) Engages in a variety of strategies to break down reading Other:
Engagement with the reading M (NI)	+ I sometimes think about how this information on plant cell structure relates to the current unit we're doing. – I do not spend time recalling what I have learned in other units.	(Demonstrates an understanding of the relationships between ideas by obtaining important information, asking questions, and defining issues or problems) Other: I will provide time at the end of class tomorrow for you to review content in unit 2. Look for ways it connects to this unit.
Application of reading M (NI)	+ I can explain how plant cells are permeable. – I rarely draw or think about diagrams while I am reading.	May apply ideas in a variety of ways including drawing, graphing, communicating, and so on Other: For tomorrow, draw me a picture that illustrates the permeability of plant cells. Review the reading as necessary.

Figure 7.1: Rubric to provide timely and effective feedback.

continued →

Success Criteria Teacher circles one: M = Mastery NI = Needs Improvement	**How Well Am I Doing?** Student identifies one strength (+) and one area for improvement (–).	**Teacher Feedback** Teacher circles or highlights successfully met criteria.
Metacognition (M) NI	+ I sometimes know when I don't understand. For example, I am still confused about how plants absorb nutrients. – I am often rushed, so I rarely stop reading so I can finish quickly. I never seek help.	(Displays ability to identify misunderstandings and monitor where reading may become confusing or difficult to understand) Identifies questions to ask to seek help Other: *It is great that you are aware of these factors! I will use some class time this week to show you how you can use whole-class and small-group discussion to seek help.*
Text variety M (NI)	+ I am better at understanding videos. – I struggle to understand the meaning of graphs in our textbook.	Understands a variety of texts including (videos,) graphs, insets, images, data tables, and so on Other: *Our small-group discussions are the perfect time to get help understanding the graphs. If your whole group is struggling, please let me know!*

Providing Timely and Effective Feedback

Feedback is essential for student growth. We're sure you remember a time when you were a student and worked really hard on an assignment or an assessment, such as a research paper or a particularly critical final exam. Either way, as soon as you handed in your work, all you could think about that day was, How did I do? The very next day, you probably walked into class just hoping that you would have an answer, any answer, to that question so that your nerves could relax!

We are not suggesting that you need to provide *immediate* feedback to students on every assignment or assessment. What matters is finding a balance in your feedback process so that you are able to prompt student reflection and growth while not getting bogged down in the infinite world of feedback. If you're not balanced in your approach, it's not hard to find yourself in a situation where you're spending more time providing feedback than the student did completing an assignment. This is not a productive feedback system! Most secondary-level students are simply not able to recognize their own errors, and if they can, they likely don't have the tools to correct their work without some form of prompting. This prompting can happen via a number of discourses, both formal and informal; however, it is critical to recognize that the longer students go without feedback, the less likely they are going to be able to reflect on that feedback and learn from it.

While some assignments, like a major research paper, will certainly require some time for you to provide feedback, there are many practical ways to deliver quick and timely feedback to students. As you saw in figure 7.1 (page 134), a well-crafted feedback tool, like a rubric, will save you quite a bit of time while also providing students with specific points to reflect on and grow. Rubrics make the end goal obvious and measurable in a concrete way while quickening the process and making feedback discrete instead of ambiguous. Having the specific success criteria they provide breaks down the larger learning targets, making it possible for learners to identify where to focus their efforts to improve overall. Likewise, when introducing a new concept, checking for understanding with a quick exit slip or entrance slip the next day and providing feedback with a simple + or – is also an efficient way of letting students know if they are on the right track.

There are limitless possibilities for effective feedback, so be creative in your feedback methods. Did you ever have a teacher who asked you and your student peers to trade papers as a means to quickly score an in-class quiz? Try this alternative: Have students keep their own quizzes and put all utensils away except for a pen that is a different color than the one they used to take the quiz. Then go over the correct answers, including the reasoning and evidence necessary to support why an answer is correct while having students adjust their own errors. Not only have you provided timely feedback, but you have also been able to help students reflect and identify misconceptions and have also avoided the shame that some students feel when exposing their work to others. The benefit of timely feedback for students is clear—less time being confused and making the same errors over and over again. As this kind of feedback makes any student error trends become clear, do this data gathering following a minilesson to refocus student learning and address those misconceptions. Even for those who demonstrated understanding on the first go, it will help to reinforce that understanding. Finally, as we cover next, it gives you data to return to your team regarding any learning trends that you identified.

Analyzing and Applying Data

After an assessment, it is critical to not only provide students with timely feedback but also bring samples of student assessments back to your collaborative team for further discussion. In terms of analysis, your assessments should be designed to gather data that are aligned with your collaborative team's learning targets and process standards. With this connection established (this occurs during the assessment-design phase), there are two major lenses (perspectives) to employ while taking a deep dive into these data: (1) the whole class and (2) a sample pack (collections of a few assessments from different student populations, such as students of certain reading levels, students with special needs, students who are learning English, and so on).

When assessment structures align with a team's learning targets, gathering data for a whole class (or even multiple classes) is very easy to do, because the data will be easy to compile. There are a number of tools that can support this process, such as Mastery Manager (www.masterymanager.com) and Google Classroom (https://classroom.google.com). These provide you with a useful way to look for overarching trends of correct and incorrect responses.

The sample-pack method works best when you want to go deeper than just looking at whole-class trends, but rather want to zoom in on the progress of more specific student groups. Thoroughly exploring a subset of assessment results takes more time, but it's an ideal approach for identifying where students may have made a wrong turn in the process, logic, or application.

When working with our team, we often employ both of these methods, first looking at final answers and scores for an entire class or population, but then also analyzing a sample pack to further identify where students made mistakes, consequently guiding our next steps to reteach. Figure 7.2 provides guiding questions for teams to conduct data analysis both for a whole class and via a sample pack.

1. Whole Class
    - As a whole, how did students perform? Are there any tasks or questions that proved problematic for the class as a whole? If yes, review how these topics or tasks were covered in your curriculum and teaching.
    - Do you see any general trends among student assessment data for multiple sections (classes) of the same course?
    - To what can you attribute student successes and challenges?
2. Sample Pack
    - What variations do you see in terms of performance by the group?
    - Is there a need for additional scaffolds for specific students demonstrating a common need?

Figure 7.2: Data-analysis perspectives.

Visit ***go.SolutionTree.com/literacy*** *for a free reproducible version of this figure.*

Once you've collected and analyzed data, the next critical step is to decide what to do with what the data are telling you. Different data points and data sets have different purposes. It may seem clear what to do after conducting an analysis of formative data, such as identifying where students need more instruction and using that knowledge to revisit a correlated learning target with a more scaffolded process. Benchmark or normative data, on the other hand, can sometimes leave a teacher wondering what exactly to do with them. In our collaborative team, we always start our discussion with the obvious—the fact that these data likely confirm or add evidence and clarification to what we have already observed in terms of student skill. If this isn't the case, then we ask, "What were the surprises?" This task can be daunting, especially if the data are confirming that you may have readers in your class operating at a four-grade (or more) deficit from the content you are teaching.

Sometimes seeing the full challenge ahead of you, on paper, may feel overwhelming. Don't let these data points thwart you; again, these data may only confirm what you probably already suspected. Even if the data come as a surprise, they represent critical information for you and your team to have. As you move forward, these data give you your marching orders for how to support your students in developing a wide array of skills. You will not be able to craft a differentiated lesson to each student in the room, nor employ multiple texts and primary sources on the same topic each day, but you can and should look for opportunities to do so throughout your unit, and science provides a great platform for this kind of work—we selected and crafted the strategies in chapters 3–6 to help you do just this.

As you deploy these strategies, consider how creating student learning groups based on assessment data can be a very productive use of your assessment data. Many online assessment systems, such as the assessment tools at Renaissance Learning (www.renaissance.com), have tools integrated as part of their reporting features that allow you to easily group students and will even generate skill data for groups. Whether using an online data tool or old-school paper and pen, here are some basic grouping methods. Which one you choose will ultimately depend on the task and purpose.

- **Homogeneous groups:** Group students based on similar assessment results (students with like scores) to focus on developing a group-specific skill or to master a group-specific content goal.
- **Linear groups:** Create a list of students from highest to lowest score and break them into groups accordingly.
- **Heterogeneous groups:** Group students with highly variable skill sets or proficiency levels so that they can benefit from each other's areas of strength. The key to this grouping is to ensure students work as a community for their mutual benefit, rather than (as we sometimes see) one or two students working while the others are either lost or copying the work without understanding it. Another way to distribute groups heterogeneously is to list the students vertically from highest score to lowest score. Then, cut the list in half, and place the two lists side-by-side. Each pair of names then signifies who works as a paired group.

While there are many more approaches your team could take, what's important is to think creatively about how you can group students in a way that allows them to grow and benefit from collaboration and collective strengths. While randomly grouping students by counting off in class or organizing based on birthdays may provide variety, these methods don't fit any specific purpose or further skill development and content mastery.

Wrapping Up

Meaningful assessment requires a great deal of intention and reflection on your part. While it is the students' responsibility to apply the skills they have developed, your role is to create assessments that intentionally and accurately mirror the curriculum and skills students must demonstrate. This kind of intentional assessment and reflection leads to responsive teaching for the entire professional learning community. This act of inquiry fosters the thoughtful selection and development of tools, texts, and tasks that help students progress toward being critically literate and possessing the disciplinary literacy skills necessary to grow not only within your content classroom but also in the post–high school world.

Collaborative Considerations *for* Teams

- Does your team have a robust assessment cycle that balances formative and summative assessments?
- Does your team take student reading ability into account when selecting texts for use in assessments? Does the team have an awareness of text and task complexity?
- Has your team collaborated on how to best share feedback with students that will impact future learning and growth?

EPILOGUE

Building capacity for collaboration among science teachers and literacy experts is truly the strongest catalyst for supporting student growth in every area of school curriculum. As schools commit to exemplary instruction to support the growth of every learner, this is something that educators with the power of a PLC behind them must not overlook. Regardless of the structure of teams in your PLC, whether teams are organized based on discipline, vertically, or across departments, this work always begins with teaming teachers who are focused on improving their own capacity to impact student learning and scaffold critical literacy skills—those that transfer to content and task.

While this book speaks directly to science experts who want to hone their teaching of literacy within their classroom, this is one installment in a series that will support teacher collaboration and strategic literacy-infused teaching in all grades and content areas. Each text in this series focuses on building a common language and literacy thought partners across all disciplines. Ultimately, the literacy skills developed in grades 6–12 students over the course of their academic career will greatly impact their ability to think critically and their overall readiness for college and career.

APPENDIX:

REPRODUCIBLES

Frayer Model Template

Directions: Place a vocabulary concept that was critical to your reading in the center square, and explain it using the included prompts.

Definition (what it is):	Characteristics:
Word:	
Examples or pictures:	What it is not:

Recalling Previous Reading to Prepare for New Learning

Directions: Recall or skim section ____________________ to help you complete the following steps.

1. In the space provided, draw an illustration of ____________________ that includes the following labels:

   ____

   ____

2. As best as you can, define ____________________ in your own words.

3. Explain how ____.

4. Explain how ____.

5. Define ____________________ in your own words.

6. Now read ____.

Activating Background Knowledge Strategy

Directions: Skim the assigned reading and note key features, such as the title, headings, images, and captions. Predict the focus (main idea) of the article and what you anticipate learning from the images and captions. Use the provided boxes to list specific features and write your initial thinking and predictions.

Focus:

Feature 1: ____________________	**Feature 2:** ____________________
Feature 3: ____________________	**Feature 4:** ____________________

Anticipation Guide

Directions:

- *Before viewing*—Look at each statement carefully. Put a check in the appropriate column in Before Viewing to indicate whether you agree or disagree with each statement.
- *During viewing*—In the center column, write the evidence from the text or video that supports or contradicts the statement. Include the location or time where you found your evidence.
- *After viewing*—Reread the statement and the evidence that supports or contradicts it. Put a check in the appropriate column in After Viewing to indicate whether you now agree or disagree.

Before Viewing		Statement and Evidence	Location	After Viewing	
Agree	Disagree			Agree	Disagree
		1. Evidence:			
		2. Evidence:			
		3. Evidence:			
		4. Evidence:			

Before Viewing		Statement and Evidence	Location	After Viewing	
Agree	Disagree			Agree	Disagree
		5. Evidence:			
		6. Evidence:			
		7. Evidence:			

Predicting and Confirming Activity

Directions: In the left-hand column, write your predictions for the assigned reading based on having previewed it. During the reading, write a plus in the center column if your prediction is confirmed, or a minus if it is refuted. In the right-hand column, record evidence from the text that supports or refutes your predictions.

Name: Text Title:		
Prediction	**(+) prediction is confirmed** **(–) prediction is not confirmed**	**Support**

Text-Dependent Questioning Graphic Organizer

Directions: Use the following generic questions to formulate your own questions about the reading that are unique to it. Your questions should reflect your own thinking as you read the text.

Science text-dependent questioning for the text ______________________________

Question Category	Questioning-the-Author (Scientist or Experiment) General Questions	My Focused Questions
Said What? What is the scientist or experiment saying?	• What is the scientist or experiment telling you? • What does the scientist or experiment say you need to clarify? • What can you do to clarify what the scientist or experiment says? • What does the scientist or experiment assume you already know?	
Did What? What did the scientist or experiment do?	• How does the scientist or experiment tell you? • Why is the scientist or experiment telling (or showing) you this fact, statistic, description, example, or visual? • What does the vocabulary reveal about the content or experiment? • How does the scientist or experiment signal what is most important? • How does the scientist construct his or her experiment or develop his or her ideas?	

Question Category	Questioning-the-Author (Scientist or Experiment) General Questions	My Focused Questions
So What? So what might the scientist or experiment mean?	• What does the scientist or experiment want you to understand? • Why is the scientist or experiment telling you this? • Does the scientist or experiment explain why something is so? • What point is the scientist or experiment making here? • What is the scientist's or experiment's purpose, and what support (evidence or reasoning) does the scientist or experiment present?	
Now What? Now, what can you do with your understanding of the scientist or experiment?	• How does this connect or apply to what I know? • How does what the scientist or experiment says influence or change your thinking? • What implications can you draw from what the scientist or experiment has told you?	

Source: Adapted from Buehl, 2017.

Simple T-Chart for Taking Notes

Directions: Use the columns to list important events or information from your reading and the reasons those events were important.

Title and chapter of text: ______________________	
List important events or information from this chapter.	**Why was this important? How does the writer feel about it? Why write about it?**

Expanded Information Chart to Extend Thinking

Directions: During the reading, complete the following graphic organizer.

What is the design element?	Describe or draw the design element.	Explain why it works. What does it resist?	Where would you use it?

Five Words Recording Sheet

Directions: While reading, underline key words and phrases. After reading, choose the five most important words from the reading, and add them to the Individual Selections column. When instructed, as a small group, discuss the words in the Individual Selections column and come to a consensus on the five most important words that the group agrees on. Add those words to the Group Consensus column.

Individual Selections	**Group Consensus**
1.	1.
2.	2.
3.	3.
4.	4.
5.	5.

Discussion Questions

Which words can your group agree on?

Which words led to disagreements?

How did your thinking change as a result of your discussion?

Self-Questioning Tool

Directions: Based on data or information your teacher has provided, use the space provided below to develop your own questions about them. (Each category offers an example of the type of question you should think of.)

Data Questions

What is the nature of the data or information we are working with?

Pattern Questions

What questions do you have that deal with patterns?

Reasoning Questions

How do the patterns (and other aspects) lead to conclusions?

Lingering Questions

What else are you wondering? Confused by?

Sample Two-Paragraph Template

Directions: Use this table as a guide to organize your sentences as you write two paragraphs on the assigned topic.

Sentence Number	Type of Sentence
1	**Topic sentence:** State your *claim* (purpose of the text).
2–3	**Lead-in:** Establish a context for the upcoming evidence, or explain a little about what you are claiming. You might define vocabulary or add context to further your idea.
4	**Quotation or evidence:** Make sure you choose one of the best quotes or pieces of evidence from the text that will support your claim.
5–6	**Justification or reasoning:** Now explain why your evidence is important. How and why does this evidence help support your claim? Explain the reasoning behind why your evidence proves your claim.

Sentence Number	Type of Sentence
7	**Quotation or evidence:** Make sure your second quote or piece of evidence supports your claim as well. Make sure you transition carefully into the evidence; don't just drop it in the middle of the paragraph with no sentence setting it up.
8–9	**Justification or reasoning:** Now explain why your evidence is important. Why does it help to support your claim? Explain the reasoning why your evidence proves your claim.
10	**Return to the claim (concluding sentence):** Return to the claim. This last sentence should relink your evidence and ideas back to your primary claim.

page 2 of 2

Evaluating Claim Statements and Supporting Them With Evidence

Directions: Three students have developed a claim to explain the figures shown below. Read each of the claims, and put a checkmark next to the student you feel has the best answer. After choosing a student's claim, read pages ____________________ in your textbook, and find at least two pieces of evidence to support your chosen student's claim.

Title of Reading: ____

Claim 1: ____________________________ says

____

____

Claim 2: ____________________________ says

____

____

Claim 3: ____________________________ says

____

____

Place or draw a figure or figures from the textbook here.

Evidence	Page Number	How the Evidence Supports Your Chosen Claim

Rubric to Provide Timely and Effective Feedback

Directions: Fill in the following information about your assignment and then assess your overall progress using the first table below. Then, use the middle column of the second table to fill in information about your strengths and areas for improvement.

Name: ______________________

Unit: ______________________ Targets: ______________________

Subject: ______________________ Literacy rubric for: ______________________

Reading unit: ______________________ Learning target: ______________________

4	3	2	1
I always read and understand a variety of texts by effectively taking notes, asking questions, and applying content. I can apply content to connect knowledge between units and across science disciplines. I am able to help my peers by teaching and evaluating their reading strategies.	I can consistently read and understand a variety of texts by effectively taking notes, asking questions, and applying the content. I recognize when I do not understand and ask specific questions to seek support.	I read and understand a variety of texts by taking notes, asking questions, or applying the content. Occasionally, I recognize when I do not understand, and I may ask questions or seek support. I make some connections between concepts within the reading.	I read for completion of a reading assignment. I cannot identify relationships or concepts within the reading. I do not observe all aspects of the text (skip pictures, graphs, and so on). I do not realize when my understanding stops, and I do not ask questions to seek support.

Success Criteria Teacher circles one: M = Mastery NI = Needs Improvement	**How Well Am I Doing?** Student identifies one strength (+) and one area for improvement (−).	**Teacher Feedback** Teacher circles or highlights successfully met criteria.
Note-taking skills M NI		Effectively uses notes to break down reading and retain information Engages in a variety of strategies to break down reading Other:
Engagement with the reading M NI		Demonstrates an understanding of the relationships between ideas by obtaining important information, asking questions, and defining issues or problems Other:
Application of reading M NI		May apply ideas in a variety of ways including drawing, graphing, communicating, and so on Other:

Success Criteria Teacher circles one: M = Mastery NI = Needs Improvement	**How Well Am I Doing?** Student identifies one strength (+) and one area for improvement (−).	**Teacher Feedback** Teacher circles or highlights successfully met criteria.
Metacognition M NI		Displays ability to identify misunderstandings and monitor where reading may become confusing or difficult to understand Identifies questions to ask to seek help Other:
Text variety M NI		Understands a variety of texts including videos, graphs, insets, images, data tables, and so on Other:

REFERENCES AND RESOURCES

Abosalem, Y. (2016, March 6). Assessment techniques and students' higher-order thinking skills. *International Journal of Secondary Education, 4*(1), 1–11. Accessed at https://pdfs.semanticscholar.org/81e4/0f2f1321180b6acf5de0d53b7f05251ba030.pdf on August 12, 2019.

Anderson, L. W., & Krathwohl, D. (Eds.). (2001). *A taxonomy for learning, teaching, and assessing: A revision of Bloom's taxonomy of educational objectives.* Boston: Allyn & Bacon.

Bloom, B. S. (1956). *Taxonomy of educational objectives, handbook I: The cognitive domain.* New York: David McKay.

Buehl, D. (2017). *Developing readers in the academic disciplines* (2nd ed.). Portland, ME: Stenhouse.

Buffum, A., Mattos, M., & Malone, J. (2018). *Taking action: A handbook for RTI at Work™.* Bloomington, IN: Solution Tree Press.

Buffum, A., Mattos, M., & Weber, C. (2012). *Simplifying response to intervention: Four essential guiding principles.* Bloomington, IN: Solution Tree Press.

Carroll, S. B. (2009). *Into the jungle: Great adventures in the search for evolution.* San Francisco: Pearson Education.

Center for Comprehensive School Reform and Improvement. (2007). *A teacher's guide to differentiating instruction.* Accessed at https://files.eric.ed.gov/fulltext/ED495740.pdf on January 9, 2019.

Columbia University Teachers College. (2005). *The academic achievement gap: Facts & figures.* Accessed at www.tc.columbia.edu/articles/2005/june/the-academic-achievement-gap-facts--figures on October 23, 2019.

Conzemius, A. E., & O'Neill, J. (2014). *The handbook for SMART school teams: Revitalizing best practices for collaboration.* Bloomington, IN: Solution Tree Press.

Csikszentmihalyi, M. (2009). *Flow: The psychology of optimal experience.* New York: HarperCollins.

Dixon, E. C., & Zannu, D. (2014). *Supporting multiple disabilities through differentiation.* Accessed at www.gadoe.org/Curriculum-Instruction-and-Assessment/Special-Education-Services/Documents/IDEAS%202014%20Handouts/

Supporting%20Multiple%20Disabilities%20through%20Differentiation %20ppt.pdf on August 14, 2018.

DuFour, R. (2004). What is a "professional learning community?" *Educational Leadership, 61*(8), 6–11. Accessed at www.siprep.org/uploaded/Professional Development/Readings/PLC.pdf on January 17, 2019.

DuFour, R., DuFour, R., Eaker, R., Many, T. W., & Mattos, M. (2016). *Learning by doing: A handbook for Professional Learning Communities at Work* (3rd ed.). Bloomington, IN: Solution Tree Press.

Gabriel, R., & Wenz, C. (2017). Three directions for disciplinary literacy. *Educational Leadership, 74*(5). Accessed at www.ascd.org/publications/educational-leadership /feb17/vol74/num05/Three-Directions-for-Disciplinary-Literacy.aspx on January 17, 2019.

Gaidos, S. (2009). Getting the dirt on carbon. *Science News for Students*. Accessed at www.sciencenewsforstudents.org/article/getting-dirt-carbon on January 17, 2019.

Gallagher, K. (2018). *Mass shooting unit: Day 5*. Accessed at www.kellygallagher.org /kellys-blog/mass-shooting-unit-day-5 on July 26, 2018.

Gambrell, L. B. (2011). Seven rules of engagement: What's most important to know about motivation to read. *The Reading Teacher, 65*(3), 172–178.

Garmston, R., & Wellman, B. (1998). Teacher talk that makes a difference. *Educational Leadership, 55*(7), 30–34.

Good, M. E. (2006). *Differentiated instruction: Principles and techniques for the elementary grades*. Accessed at https://eric.ed.gov/?id=ED491580 on January 9, 2019.

Harris, T., & Threewitt, C. (n.d.). *How roller coasters work*. Accessed at https://science .howstuffworks.com/engineering/structural/roller-coaster3.htm on September 30, 2019.

International Society for Technology in Education. (n.d.). *ISTE standards for students*. Accessed at www.iste.org/standards/for-students on September 3, 2019.

Ketcham, C. (2012, Winter). Warning: High frequency. *Earth Island Journal*. Accessed at www.earthisland.org/journal/index.php/magazine/entry/warning_high_frequency on January 17, 2019.

Lorant, G. (2016). *Seismic design principles*. Accessed at www.wbdg.org/resources /seismic-design-principles on August 9, 2019.

McFarland, J., Hussar, B., Wang, X., Zhang, J., Wang, K., Rathbun, A., et al. (2018). *The condition of education 2018*. Accessed at https://nces.ed.gov/pubs2018/2018144 .pdf on May 29, 2019.

McGlynn, K., & Kozlowski, J. (2017). Science for all: Kinesthetic learning in science. *Science Scope, 40*(9), 24–28.

McKnight, K. S. (2010). *The teacher's big book of graphic organizers: 100 reproducible organizers that help kids with reading, writing, and the content areas*. San Francisco: Jossey-Bass.

Mitchell, D. (2010). *Education that fits: Review of international trends in the education of students with special educational needs—Chapter 11: Inclusive education.* Christchurch, New Zealand: University of Canterbury. Accessed at www.educationcounts.govt.nz/publications/special_education/education-that-fits/chapter-eleven-inclusive-education on August 14, 2018.

National Center on Intensive Intervention. (n.d.). *Intensive intervention & multi-tiered systems of support (MTSS).* Accessed at https://intensiveintervention.org/intensive-intervention/multi-tiered-systems-support on September 4, 2019.

National Governors Association Center for Best Practices & Council of Chief State School Officers. (2010). *Common Core State Standards for English language arts and literacy in history/social studies, science, and technical subjects.* Washington, DC: Authors. Accessed at www.corestandards.org/assets/CCSSI_ELA%20Standards.pdf on July 24, 2018.

NGSS Lead States. (2013). *Next Generation Science Standards: For states, by states.* Washington, DC: The National Academies Press. Accessed at www.nextgenscience.org/search-standards on May 6, 2019.

Pardede, P. (2017). *A review on reading theories and its implication to the teaching of reading.* Paper presented at the Bimonthly Collegiate Forum of Universitas Kristen Indonesia, Jakarta, Indonesia.

The Physics Classroom. (n.d.). *What is a projectile?* Accessed at www.physicsclassroom.com/class/vectors/Lesson-2/What-is-a-Projectile on August 9, 2019.

Rebell, M. A. (2008, February). Equal opportunity and the courts. *Phi Delta Kappan, 89*(6), 432–439.

RTI Action Network. (n.d.). *Tiered instruction/intervention.* Accessed at http://rtinetwork.org/essential/tieredinstruction on January 18, 2019.

Rumelhart, D. E. (1980). Schemata: The building blocks of cognition. In R. J. Spiro, B. C. Bruce, & W. F. Brewer (Eds.), *Theoretical issues in reading comprehension: Perspective and cognitive psychology, linguistics, artificial intelligence, and education* (pp. 33–58). Hillsdale, NJ: Erlbaum.

Shanahan, T., & Shanahan, C. (2008). Teaching disciplinary literacy to adolescents: Rethinking content-area literacy. *Harvard Educational Review, 78*(1), 40–59, 279. Accessed at https://dpi.wi.gov/sites/default/files/imce/cal/pdf/teaching-dl.pdf on January 17, 2019.

ThinkCERCA. (n.d.a). *Our story.* Accessed at https://thinkcerca.com/about/our-story on July 24, 2018.

ThinkCERCA. (n.d.b). *Why argumentation?: Our research-based approach.* Accessed at https://thinkcerca.com/argumentation-research-based-approach on January 18, 2019.

Tovani, C. (2000). *I read it, but I don't get it: Comprehension strategies for adolescent readers.* Portland, ME: Stenhouse.

University of New Hampshire. (n.d.). *An introduction to the global carbon cycle.* Accessed at http://globecarboncycle.unh.edu/CarbonCycleBackground.pdf on September 4, 2019.

U.S. Department of Education, National Center for Education Statistics, National Assessment of Educational Progress. (n.d.). *NAEP reading report card.* Accessed at www.nationsreportcard.gov/reading_2017?grade=4 on September 4, 2019.

U.S. Department of Education, National Center for Education Statistics, National Assessment of Educational Progress. (2016). *Reading performance.* Accessed at https://nces.ed.gov/programs/coe/pdf/Indicator_CNB/coe_cnb_2016_05.pdf on March 12, 2019.

Worthy, J., & Broaddus, K. (2001). Fluency beyond the primary grades: From group performance to silent, independent reading. *Reading Teacher*, *55*(4), 334–343.

INDEX

A

B

C

D

E

P

Q

R

S

W